T0325078

Queueing Modelling Fundamentals

Queueing Modelling Fundamentals

With Applications in Communication Networks

Second Edition

Ng Chee-Hock and Soong Boon-Hee
Both of Nanyang Technological University, Singapore

John Wiley & Sons, Ltd

Other Wiley Editorial Offices

John Wiley & Sons Inc., 111 River Street, Hoboken, NJ 07030, USA

Jossey-Bass, 989 Market Street, San Francisco, CA 94103-1741, USA

Wiley-VCH Verlag GmbH, Boschstr. 12, D-69469 Weinheim, Germany

John Wiley & Sons Australia Ltd, 42 McDougall Street, Milton, Queensland 4064, Australia

John Wiley & Sons (Asia) Pte Ltd, 2 Clementi Loop #02-01, Jin Xing Distripark, Singapore 129809

John Wiley & Sons Canada Ltd, 6045 Freemont Blvd, Mississauga, ONT, L5R 4J3, Canada

Wiley also publishes its books in a variety of electronic formats. Some content that appears in print may not be available in electronic books.

Library of Congress Cataloging-in-Publication Data

Ng, Chee-Hock.
    Queueing modelling fundamentals with applications in communication networks /
Chee-Hock Ng and Boon-Hee Song. – 2nd ed.
      p. cm.
    Includes bibliographical references and index.
    ISBN 978-0-470-51957-8 (cloth)
  1. Queueing theory.   2. Telecommunication–Traffic.   I. Title.
    QA274.8.N48 2008
    519.8′2 – dc22

                                              2008002732

British Library Cataloguing in Publication Data

A catalogue record for this book is available from the British Library

ISBN 978-0-470-51957-8 (HB)

Typeset by SNP Best-set Typesetter Ltd., Hong Kong
Printed and bound in Great Britain by TJ International, Padstow, Cornwall

*To my wife, Joyce,
and three adorable children, Sandra,
Shaun and Sarah, with love*
—NCH

*To my wonderful wife, Buang Eng and
children Jareth, Alec and Gayle,
for their understanding*
—SBH

Contents

List of Tables

List of Illustrations

Preface

Welcome to the second edition of *Queueing Modelling Fundamentals With Applications in Communication Networks*. Since the publication of the first edition by the first author Ng Chee-Hock in 1996, this book has been adopted for use by several colleges and universities. It has also been used by many professional bodies and practitioners for self-study.

This second edition is a collaborative effort with the coming on board of a second author to further expand and enhance the contents of the first edition. We have in this edition thoroughly revised all the chapters, updated examples and problems included in the text and added more worked examples and performance curves. We have included new materials/sections in several of the chapters, as well as a new Chapter 9 on 'Flow and Congestion Control' to further illustrate the various applications of queueing theory. A section on 'Network Calculus' is also added to Chapter 8 to introduce readers to a set of recent developments in queueing theory that enables deterministic bounds to be derived.

INTENDED AUDIENCE

Queueing theory is often taught at the senior level of an undergraduate programme or at the entry level of a postgraduate course in computer networking or engineering. It is often a prerequisite to some more advanced courses such as network design and capacity planning.

This book is intended as an introductory text on queueing modelling with examples on its applications in computer networking. It focuses on those queueing modelling techniques that are useful and applicable to the study of data networks and gives an in-depth insight into the underlying principles of isolated queueing systems as well as queueing networks. Although a great deal of effort is spent in discussing the models, their general applications are demonstrated through many worked examples.

It is the belief of the authors and experience learned from many years of teaching that students generally absorb the subject matter faster if the underlying concepts are demonstrated through examples. This book contains many worked examples intended to supplement the teaching by illustrating the possible applications of queueing theory. The inclusion of a large number of examples aims to strike a balance between theoretical treatment and practical applications.

This book assumes that students have a prerequisite knowledge on probability theory, transform theory and matrices. The mathematics used is appropriate for those students in computer networking and engineering. The detailed step-by-step derivation of queueing results makes it an excellent text for academic courses, as well as a text for self-study.

ORGANISATION OF THIS BOOK

This book is organised into nine chapters as outlined below:

Chapter 1 refreshes the memory of students on those mathematical tools that are necessary prerequisites. It highlights some important results that are crucial to the subsequent treatment of queueing systems.

Chapter 2 gives an anatomy of a queueing system, the various random variables involved and the relationships between them. It takes a close look at a frequently-used arrival process – the Poisson process and its stochastic properties.

Chapter 3 introduces Markov processes that play a central role in the analysis of all the basic queueing systems – Markovian queueing systems.

Chapter 4 considers single-queue Markovian systems with worked examples of their applications. Emphasis is placed on the techniques used to derive the performance measures for those models that are widely used in computer communications and networking. An exhaustive listing of queueing models is not intended.

Chapter 5 looks at semi-Markovian systems, M/G/1 and its variants. G/M/1 is also mentioned briefly to contrast the random observer property of these two apparently similar but conceptually very different systems.

Chapter 6 extends the analysis to the open queueing networks with a single class of customers. It begins with the treatment of two tandem queues and its limitations in applying the model to transmission channels in series, and subsequently introduces the Jackson queueing networks.

Chapter 7 completes the analysis of queueing networks by looking at other types of queueing networks – closed queueing networks. Again treatments are limited to the networks with a single class of customers.

Chapter 8 looks at some more exotic classes of arrival processes used to model those arrivals by correlation, namely the Markov-modulated Poisson process, the Markov-modulated Bernoulli process and the Markov-modulated fluid flow. It also briefly introduces a new paradigm of deterministic queueing called network calculus that allows deterministic bounds to be derived.

Chapter 9 looks at the traffic situation in communication networks where queueing networks can be applied to study the performance of flow control mechanisms. It also briefly introduces the concept of sliding window and rate-based flow control mechanisms. Finally, several buffer allocation schemes are studied using Markovian systems that combat congested states.

ACKNOWLEDGEMENTS

This book would not have been possible without the support and encouragement of many people. We are indebted to Tan Chee Heng, a former colleague, for painstakingly going through the manuscripts of the first edition.

The feedback and input of students who attended our course, who used this book as the course text, have also helped greatly in clarifying the topics and examples as well as improving the flow and presentation of materials in this edition of the book. Finally, we are grateful to Sarah Hinton, our Project Editor and Mark Hammond at John Wiley & Sons, Ltd for their enthusiastic help and patience.

1

Preliminaries

Queueing theory is an intricate and yet highly practical field of mathematical study that has vast applications in performance evaluation. It is a subject usually taught at the advanced stage of an undergraduate programme or the entry level of a postgraduate course in Computer Science or Engineering. To fully understand and grasp the essence of the subject, students need to have certain background knowledge of other related disciplines, such as probability theory and transform theory, as a prerequisite.

It is not the intention of this chapter to give a fine exposition of each of the related subjects but rather meant to serve as a refresher and highlight some basic concepts and important results in those related topics. These basic concepts and results are instrumental to the understanding of queueing theory that is outlined in the following chapters of the book. For more detailed treatment of each subject, students are directed to some excellent texts listed in the references.

1.1 PROBABILITY THEORY

In the study of a queueing system, we are presented with a very dynamic picture of events happening within the system in an apparently random fashion. Neither do we have any knowledge about when these events will occur nor are we able to predict their future developments with certainty. Mathematical models have to be built and probability distributions used to quantify certain parameters in order to render the analysis mathematically tractable. The

Queueing Modelling Fundamentals   Second Edition   Ng Chee-Hock and Soong Boon-Hee
© 2008 John Wiley & Sons, Ltd

importance of probability theory in queueing analysis cannot be over-emphasized. It plays a central role as that of the limiting concept to calculus. The development of probability theory is closely related to describing randomly occurring events and has its roots in predicting the random outcome of playing games. We shall begin by defining the notion of an event and the sample space of a mathematical experiment which is supposed to mirror a real-life phenomenon.

1.1.1   Sample Spaces and Axioms of Probability

A *sample space* (Ω) of a random experiment is a collection of all the mutually exclusive and exhaustive simple outcomes of that experiment. A particular simple outcome (ω) of an experiment is often referred to as a *sample point*. An *event* (E) is simply a subset of Ω and it contains a set of sample points that satisfy certain common criteria. For example, an event could be the even numbers in the toss of a dice and it contains those sample points $\{[2], [4], [6]\}$. We indicate that the outcome ω is a sample point of an event E by writing $\{\omega \in E\}$. If an event E contains no sample points, then it is a null event and we write $E = \varnothing$. Two events E and F are said to be mutually exclusive if they have no sample points in common, or in other words the intersection of events E and F is a null event, i.e. $E \cap F = \varnothing$.

There are several notions of probability. One of the classic definitions is based on the relative frequency approach in which the probability of an event E is the limiting value of the proportion of times that E was observed. That is

$$P(E) = \lim_{N \to \infty} \frac{N_E}{N} \qquad (1.1)$$

where N_E is the number of times event E was observed and N is the total number of observations. Another one is the so-called axiomatic approach where the probability of an event E is taken to be a real-value function defined on the family of events of a sample space and satisfies the following conditions:

Axioms of probability

(i) $0 \leq P(E) \leq 1$ for any event in that experiment
(ii) $P(\Omega) = 1$
(iii) If E and F are mutually exclusive events, i.e. $E \in F = \varnothing$, then $P(E \cup F)$
      $= P(E) + P(F)$

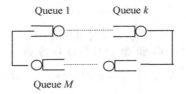

Figure 1.1   A closed loop of M queues

There are some fundamental results that can be deduced from this axiomatic definition of probability and we summarize them without proofs in the following propositions.

Proposition 1.1

(i) $P(\varnothing) = 0$
(ii) $P(\bar{E}) + P(E) = 1$ for any event E in Ω, where $\bar{E}$ is the compliment of E.
(iii) $P(E \cup F) = P(E) + P(F) - P(E \cap F)$, for any events E and F.
(iv) $P(E) \leq P(F)$, if $E \subseteq F$.
(v) $P\left(\bigcup_i E_i\right) = \sum_i P(E_i)$, for $E_i \cap E_j = \varnothing$, when $i \neq j$.

Example 1.1

By considering the situation where we have a closed loop of M identical queues, as shown in Figure 1.1, then calculate the probability that Queue 1 is non-empty (it has at least one customer) if there are N customers circulating among these queues.

Solution

To calculate the required probability, we need to find the total number of ways of distributing those N customers among M queues. Let $X_i(>0)$ be the number of customers in Queue i, then we have

$$X_1 + X_2 + \ldots + X_M = N$$

The problem can now be formulated by having these N customers lined up together with M imaginary zeros, and then dividing them into M groups. These M zeros are introduced so that we may have empty queues. They also ensure that one of the queues will contain all the customers, even in the case where

Queue dividing points

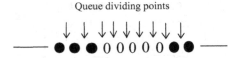

Figure 1.2 N customers and M zeros, $(N + M - 1)$ spaces

all zeros are consecutive because there are only $(M - 1)$ spaces among them, as shown in Figure 1.2.

We can select $M - 1$ of the $(N + M - 1)$ spaces between customers as our separating points and hence the number of combinations is given by

$$\binom{N+M-1}{M-1} = \binom{N+M-1}{N}.$$

When Queue 1 is empty, the total number of ways of distributing N customers among $(M - 1)$ queues is given by

$$\binom{N+M-2}{N}.$$

Therefore, the probability that Queue 1 is non-empty:

$$= 1 - \binom{N+M-2}{N} \Big/ \binom{N+M-1}{N}$$

$$= 1 - \frac{M-1}{N+M-1} = \frac{N}{N+M-1}$$

Example 1.2

Let us suppose a tourist guide likes to gamble with his passengers as he guides them around the city on a bus. On every trip, there are about 50 random passengers. Each time he challenges his passengers by betting that if there is at least two people on the bus that have the same birthday, then all of them would have to pay him $1 each. However, if there were none for that group on that day, he would repay each of them $1. What is the likelihood (or probability) of the event that he wins his bet?

Solution

Let us assume that each passenger is equally likely to have their birthday on any day of the year (we will neglect leap years). In order to solve this problem we need to find the probability that nobody on that bus has the same birthday. Imagine that we line up these 50 passengers, and the first passenger has 365 days to choose as his/her birthday. The next passenger has the remainder of 364 days to choose from in order for him/her not to have the same birthday as the first person (i.e. he has a probability of 364/365). This number of choices reduces until the last passenger. Therefore:

$$P(\text{None of the 50 passengers has the same birthday}) =$$
$$\underbrace{\left(\frac{364}{365}\right)\left(\frac{363}{365}\right)\left(\frac{362}{365}\right)\cdots\left(\frac{365-49}{365}\right)}_{49\,terms}$$

Therefore, the probability that the tourist guide wins his bet can be obtained by Proposition 1.1 (ii):

$$P(\text{At least 2 passengers out of 50 has the same birthday}) =$$
$$1 - \left(\frac{\prod_{j=1}^{49}(365-j)}{365^{49}}\right) = 0.9704.$$

The odds are very much to the favour of tourist guide, although we should remember this probability has a limiting value of (1.1) only.

1.1.2   Conditional Probability and Independence

In many practical situations, we often do not have information about the outcome of an event but rather information about related events. Is it possible to infer the probability of an event using the knowledge that we have about these other events? This leads us to the idea of *conditional probability* that allows us to do just that!

Conditional probability that an event E occurs, given that another event F has already occurred, denoted by $P(E\,|\,F)$, is defined as

$$P(E|F) = \frac{P(E \cap F)}{P(F)} \quad \text{where} \quad P(F) \neq 0 \tag{1.2}$$

Conditional probability satisfies the axioms of probability and is a probability measure in the sense of those axioms. Therefore, we can apply any results obtained for a normal probability to a conditional probability. A very useful expression, frequently used in conjunction with the conditional probability, is the so-called *Law of Total Probability*. It says that if $\{A_i \in \Omega, i = 1, 2, \ldots, n\}$ are events such that

(i) $A_i \cap A_j = \emptyset$ if $i \neq j$

(ii) $P(A_i) > 0$

(iii) $\bigcup_{i=1}^{n} A_i = \Omega$

then for any event E in the same sample space:

$$P(E) = \sum_{i=1}^{n} P(E \cap A_i) = \sum_{i=1}^{n} P(E|A_i)P(A_i) \tag{1.3}$$

This particular law is very useful for determining the probability of a complex event E by first conditioning it on a set of simpler events $\{Ai\}$ and then by summing up all the conditional probabilities. By substituting the expression (1.3) in the previous expression of conditional probability (1.2), we have the well-known *Bayes' formula*:

$$P(E|F) = \frac{P(E \cap F)}{\sum_i P(F|A_i)P(A_i)} = \frac{P(F|E)P(E)}{\sum_i P(F|A_i)P(A_i)} \tag{1.4}$$

Two events are said to be statistically independent if and only if

$$P(E \cap F) = P(E)P(F).$$

From the definition of conditional probability, this also implies that

$$P(E|F) = \frac{P(E \cap F)}{P(F)} = \frac{P(E)P(F)}{P(F)} = P(E) \tag{1.5}$$

Students should note that the statistical independence of two events E and F does not imply that they are mutually exclusive. If two events are mutually exclusive then their intersection is a null event and we have

$$P(E|F) = \frac{P(E \cap F)}{P(F)} = 0 \quad \text{where} \quad P(F) \neq 0 \tag{1.6}$$

Example 1.3

Consider a switching node with three outgoing links A, B and C. Messages arriving at the node can be transmitted over one of them with equal probability. The three outgoing links are operating at different speeds and hence message transmission times are 1, 2 and 3 ms, respectively for A, B and C. Owing to the difference in trucking routes, the probability of transmission errors are 0.2, 0.3 and 0.1, respectively for A, B and C. Calculate the probability of a message being transmitted correctly in 2 ms.

Solution

Denote the event that a message is transmitted correctly by F, then we are given

$$P(F \mid A\ Link) = 1 - 0.2 = 0.8$$
$$P(F \mid B\ Link) = 1 - 0.3 = 0.7$$
$$P(F \mid C\ Link) = 1 - 0.1 = 0.9$$

The probability that a message being transmitted correctly in 2 ms is simply the event $(F \cap B)$, hence we have

$$P(F \cap B) = P(F|B) \times P(B)$$
$$= 0.7 \times \frac{1}{3} = \frac{7}{30}$$

1.1.3   Random Variables and Distributions

In many situations, we are interested in some numerical value that is associated with the outcomes of an experiment rather than the actual outcomes themselves. For example, in an experiment of throwing two die, we may be interested in the sum of the numbers (X) shown on the dice, say $X = 5$. Thus we are interested in a function which maps the outcomes onto some points or an interval on the real line. In this example, the outcomes are $\{2,3\}$, $\{3,2\}$, $\{1,4\}$ and $\{4,1\}$, and the point on the real line is 5.

This mapping (or function) that assigns a real value to each outcome in the sample space is called a *random variable*. If X is a random variable and x is a real number, we usually write $\{X \leq x\}$ to denote the event $\{\omega \in \Omega$ and $X(\omega) \leq x\}$. There are basically two types of random variables; namely the *discrete random variables* and *continuous random variables*. If the mapping function assigns a real number, which is a point in a countable set of points on the real line, to an outcome then we have a discrete random variable. On the other hand, a continuous random variable takes on a real number which falls in an interval on the real line. In other words, a discrete random variable can assume at most a finite or a countable infinite number of possible values and a continuous random variable can assume any value in an interval or intervals of real numbers.

A concept closely related to a random variable is its *cumulative probability distribution function*, or just *distribution function* (*PDF*). It is defined as

$$F_X(x) \equiv P[X \leq x]$$
$$= P[\omega: X(\omega) \leq x] \tag{1.7}$$

For simplicity, we usually drop the subscript X when the random variable of the function referred to is clear in the context. Students should note that a distribution function completely describes a random variable, as all parameters of interest can be derived from it. It can be shown from the basic axioms of probability that a distribution function possesses the following properties:

Proposition 1.2

(i) F is a non-negative and non-decreasing function, i.e. if $x_1 \leq x_2$ then $F(x_1) \leq F(x_2)$
(ii) $F(+\infty) = 1$ & $F(-\infty) = 0$
(iii) $F(b) - F(a) = P[a < X \leq b]$

For a discrete random variable, its probability distribution function is a disjoint step function, as shown in Figure 1.3. The probability that the random variable takes on a particular value, say x and $x = 0, 1, 2, 3 \ldots$, is given by

$$p(x) \equiv P[X = x] = P[X < x+1] - P[X < x]$$
$$= \{1 - P[X \geq x+1]\} - \{1 - P[X \geq x]\}$$
$$= P[X \geq x] - P[X \geq x+1] \tag{1.8}$$

The above function $p(x)$ is known as the *probability mass function* (*pmf*) of a discrete random variable X and it follows the axiom of probability that

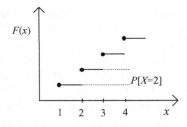

Figure 1.3   Distribution function of a discrete random variable X

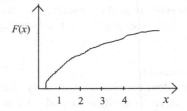

Figure 1.4   Distribution function of a continuous RV

$$\sum_x p(x) = 1 \tag{1.9}$$

This probability mass function is a more convenient form to manipulate than the *PDF* for a discrete random variable.

In the case of a continuous random variable, the probability distribution function is a continuous function, as shown in Figure 1.4, and *pmf* loses its meaning as $P[X = x] = 0$ for all real x.

A new useful function derived from the *PDF* that completely characterizes a continuous random variable X is the so-called *probability density function* (*pdf*) defined as

$$f_X(x) \equiv \frac{d}{dx} F_X(x) \tag{1.10}$$

It follows from the **axioms of probability** and the definition of *pdf* that

$$F_X(x) = \int_{-\infty}^{x} f_X(\tau) d\tau \tag{1.11}$$

$$P[a \le X \le b] = \int_a^b f_X(x) dx \tag{1.12}$$

and

$$\int_{-\infty}^{\infty} f_X(x) = P[-\infty < X < \infty] = 1 \tag{1.13}$$

The expressions (1.9) and (1.13) are known as the *normalization conditions* for discrete random variables and continuous random variables, respectively.

We list in this section some important discrete and continuous random variables which we will encounter frequently in our subsequent studies of queueing models.

(i) Bernoulli random variable
A *Bernoulli trial* is a random experiment with only two outcomes, 'success' and 'failure', with respective probabilities, p and q. A *Bernoulli random variable X* describes a Bernoulli trial and assumes only two values: 1 (for success) with probability p and 0 (for failure) with probability q:

$$P[X = 1] = p \quad \& \quad P[X = 0] = q = 1 - p \tag{1.14}$$

(ii) Binomial random variable
If a Bernoulli trial is repeated k times then the random variable X that counts the number of successes in the k trials is called a *binomial random variable* with parameters k and p. The probability mass function of a binomial random variable is given by

$$B(k; n, p) = \binom{n}{k} p^k q^{n-k} \quad k = 0, 1, 2, \ldots, n \quad \& \quad q = 1 - p \tag{1.15}$$

(iii) Geometric random variable
In a sequence of independent Bernoulli trials, the random variable X that counts the number of trials up to and including the first success is called a *geometric random variable* with the following *pmf*:

$$P[X = k] = (1 - p)^{k-1} p \quad k = 1, 2, \ldots \infty \tag{1.16}$$

(iv) Poisson random variable
A random variable X is said to be *Poisson random variable* with parameter λ if it has the following mass function:

$$P[X = k] = \frac{\lambda^k}{k!} e^{-\lambda} \quad k = 0, 1, 2, \ldots \tag{1.17}$$

Students should note that in subsequent chapters, the Poisson mass function is written as

$$P[X = k] = \frac{(\lambda' t)^k}{k!} e^{-\lambda' t} \quad k = 0, 1, 2, \ldots \tag{1.18}$$

Here, the λ in expression (1.17) is equal to the $\lambda' t$ in expression (1.18).

(v) Continuous uniform random variable

A continuous random variable X with its probabilities distributed uniformly over an interval (a, b) is said to be a *uniform random variable* and its density function is given by

$$f(x) = \begin{cases} \dfrac{1}{b-a} & a < x < b \\ 0 & otherwise \end{cases} \qquad (1.19)$$

The corresponding distribution function can be easily calculated by using expression (1.11) as

$$F(x) = \begin{cases} 0 & x < a \\ \dfrac{x-a}{b-a} & a \le x < b \\ 1 & x \ge b \end{cases} \qquad (1.20)$$

(vi) Exponential random variable

A continuous random variable X is an *exponential random variable* with parameter $\lambda > 0$, if its density function is defined by

$$f(x) = \begin{cases} \lambda e^{-\lambda x} & x > 0 \\ 0 & x \le 0 \end{cases} \qquad (1.21)$$

The distribution function is then given by

$$F(x) = \begin{cases} 1 - \lambda e^{-\lambda x} & x > 0 \\ 0 & x \le 0 \end{cases} \qquad (1.22)$$

(vii) Gamma random variable

A continuous random variable X is said to have a gamma distribution with parameters $\alpha > 0$ and $\lambda > 0$, if its density function is given by

$$f(x) = \begin{cases} \dfrac{\lambda^{\alpha}(x)^{\alpha-1} e^{-\lambda x}}{\Gamma(\alpha)} & x > 0 \\ 0 & x \le 0 \end{cases} \qquad (1.23)$$

where $\Gamma(\alpha)$ is the gamma function defined by

$$\Gamma(\alpha) = \int_{0}^{\infty} x^{\alpha-1} e^{-x} dx \quad \alpha > 0 \qquad (1.24)$$

There are certain nice properties about gamma functions, such as

$$\Gamma(k) = (k-1)\Gamma(k-1) = (k-1)! \alpha = n \text{ a positive integer}$$

$$\Gamma(\alpha) = \alpha\Gamma(\alpha-1) \quad \alpha > 0 \text{ a real number} \tag{1.25}$$

(viii) Erlang-k or k-stage Erlang Random Variable
This is a special case of the gamma random variable when α ($=k$) is a positive integer. Its density function is given by

$$f(x) = \begin{cases} \dfrac{\lambda^k (x)^{k-1}}{(k-1)!} e^{-\lambda x} & x > 0 \\[2mm] 0 & x \le 0 \end{cases} \tag{1.26}$$

(ix) Normal (Gaussian) Random Variable
A frequently encountered continuous random variable is the Gaussian or Normal with the parameters of μ (mean) and σ_X (standard deviation). It has a density function given by

$$f_X(x) = \frac{1}{\sqrt{2\pi\sigma_X^2}} e^{-(x-\mu)^2/2\sigma_X^2} \tag{1.27}$$

The normal distribution is often denoted in a short form as $N(\mu, \sigma_X^2)$.

Most of the examples above can be roughly separated into either *continuous* or *discrete* random variables. A discrete random variable can take on only a finite number of values in any finite observations (e.g. the number of heads obtained in throwing 2 independent coins). On the other hand, a continuous random variable can take on any value in the observation interval (e.g. the time duration of telephone calls). However, samples may exist, as we shall see later, where the random variable of interest is a *mixed* random variable, i.e. they have both continuous and discrete portions. For example, the waiting time distribution function of a queue in Section 4.3 can be shown as

$$F_W(t) = (1 - \rho e^{-\mu(1-\rho)t}) \quad t \ge 0$$
$$= 0 \qquad\qquad\qquad t < 0.$$

This has a discrete portion that has a jump at $t = 0$ but with a continuous portion elsewhere.

1.1.4   Expected Values and Variances

As discussed in Section 1.1.3, the distribution function or *pmf* (*pdf*, in the case of continuous random variables) provides a complete description of a random variable. However, we are also often interested in certain measures which summarize the properties of a random variable succinctly. In fact, often these are the only parameters that we can observe about a random variable in real-life problems.

The most important and useful measures of a random variable X are its *expected value*[1] $E[X]$ and *variance Var*[X]. The expected value is also known as the *mean value* or *average value*. It gives the average value taken by a random variable and is defined as

$$E[X] = \sum_{k=0}^{\infty} kP[X = k] \quad \text{for discrete variables} \tag{1.28}$$

and

$$E[X] = \int_{0}^{\infty} xf(x)dx \quad \text{for continuous variables} \tag{1.29}$$

The variance is given by the following expressions. It measures the dispersion of a random variable X about its mean $E[X]$:

$$\begin{aligned}
\sigma^2 &= Var[X] \\
&= E[(X - E[X])^2] \quad \text{for discrete variables} \\
&= E[X^2] - (E[X])^2
\end{aligned} \tag{1.30}$$

$$\begin{aligned}
\sigma^2 &= Var[X] \\
&= \int_{0}^{\infty} (x - E[X])^2 f(x)dx \quad \text{for continuous variables} \\
&= \int_{0}^{\infty} x^2 f(x)dx - 2E[X]\int_{0}^{\infty} xf(x)dx + \int_{0}^{\infty} f(x)dx \\
&= E[X^2] - (E[X])^2
\end{aligned} \tag{1.31}$$

σ refers to the square root of the variance and is given the special name of *standard deviation*.

Example 1.4

For a discrete random variable X, show that its expected value is also given by

$$E[X] = \sum_{k=0}^{\infty} P[X > k]$$

[1] For simplicity, we assume here that the random variables are non-negative.

Solution

By definition, the expected value of X is given by

$$E[X] = \sum_{k=0}^{\infty} kP[X=k] = \sum_{k=0}^{\infty} k\{P[X \geq k] - P[X \geq k+1]\} \quad \text{(see (1.8))}$$

$$= P[X \geq 1] - [X \geq 2] + 2P[X \geq 2] - 2P[X \geq 3]$$
$$+ 3P[X \geq 3] - 3P[X \geq 4] + 4P[X \geq 4] - 4P[X \geq 5] + \ldots$$

$$= \sum_{k=1}^{\infty} P[X \geq k] = \sum_{k=0}^{\infty} P[X > k]$$

Example 1.5

Calculate the expected values for the Binomial and Poisson random variables.

Solution

1. Binomial random variable

$$E[X] = \sum_{k=1}^{n} k \binom{n}{k} p^k (1-p)^{n-k}$$

$$= np \sum_{k=1}^{n} \binom{n-1}{k-1} p^{k-1} (1-P)^{n-k}$$

$$= np \sum_{j=0}^{n} \binom{n-1}{j} p^j (1-p)^{(n-1)-j}$$

$$= np$$

2. Poisson random variable

$$E[X] = \sum_{k=0}^{\infty} k \frac{\lambda^k}{k!} e^{-\lambda}$$

$$= e^{-\lambda}(\lambda) \sum_{k=1}^{\infty} \frac{(\lambda)^{k-1}}{(k-1)!}$$

$$= e^{-\lambda}(\lambda) e^{\lambda}$$

$$= \lambda$$

Table 1.1 Means and variances of some common random variables

Random variable	$E[X]$	$Var[X]$
Bernoulli	p	pq
Binomial	np	npq
Geometric	$1/p$	q/p^2
Poisson	λ	λ
Continuous uniform	$(a+b)/2$	$(b-a)^2/12$
Exponential	$1/\lambda$	$1/\lambda^2$
Gamma	α/λ	α/λ^2
Erlang-k	$1/\lambda$	$1/k\lambda^2$
Gaussian	μ	σ_X^2

Table 1.1 summarizes the expected values and variances for those random variables discussed earlier.

Example 1.6

Find the expected value of a Cauchy random variable X, where the density function is defined as

$$f(x) = \frac{1}{\pi(1+x^2)} u(x)$$

where $u(x)$ is the unit step function.

Solution

Unfortunately, the expected value of $E[X]$ in this case is

$$E[X] = \int_{-\infty}^{\infty} \frac{x}{\pi(1+x^2)} u(x) dx = \int_{0}^{\infty} \frac{x}{\pi(1+x^2)} dx = \infty$$

Sometimes we get unusual results with expected values, even though the distribution of the random variable is well behaved.

Another useful measure regarding a random variable is the coefficient of variation which is the ratio of standard deviation to the mean of that random variable:

$$C_x \equiv \frac{\sigma_X}{E[X]}$$

1.1.5   Joint Random Variables and Their Distributions

In many applications, we need to investigate the joint effect and relationships between two or more random variables. In this case we have the natural extension of the distribution function to two random variables X and Y, namely the *joint distribution function*. Given two random variables X and Y, their joint distribution function is defined as

$$F_{XY}(x, y) \equiv P[X \leq x, Y \leq y] \qquad (1.32)$$

where x and y are two real numbers. The individual distribution function F_X and F_Y, often referred to as the *marginal distribution* of X and Y, can be expressed in terms of the joint distribution function as

$$F_X(x) = F_{XY}(x, \infty) = P[X \leq x, Y \leq \infty]$$
$$F_Y(y) = F_{XY}(\infty, y) = P[X \leq \infty, Y \leq y] \qquad (1.33)$$

Similar to the one-dimensional case, the joint distribution also enjoys the following properties:

(i) $F_{XY}(-\infty, y) = F_{XY}(x, -\infty) = 0$
(ii) $F_{XY}(-\infty, -\infty) = 0$ and $F_{XY}(\infty, \infty) = 1$
(iii) $F_{XY}(x_1, y) \leq F_{XY}(x_2, y)$ *for* $x_1 \leq x_2$
(iv) $F_{XY}(x, y_1) \leq F_{XY}(x, y_2)$ *for* $y_1 \leq y_2$
(v) $P[x_1 < X \leq x_2, y_1 < Y \leq y_2] = F_{XY}(x_2, y_2) - F_{XY}(x_1, y_2)$
$$\qquad\qquad\qquad\qquad\qquad - F_{XY}(x_2, y_1) + F_{XY}(x_1, y_1)$$

If both X and Y are jointly continuous, we have the associated *joint density function* defined as

$$f_{XY}(x, y) \equiv \frac{d^2}{dxdy} F_{XY}(x, y) \qquad (1.34)$$

and the marginal density functions and joint probability distribution can be computed by integrating over all possible values of the appropriate variables:

$$f_X(x) = \int_{-\infty}^{\infty} f_{XY}(x, y)dy$$

$$f_Y(y) = \int_{-\infty}^{\infty} f_{XY}(x, y)dx \qquad (1.35)$$

$$F_{XY}(x, y) = \int\limits_{-\infty}^{x} \int\limits_{-\infty}^{y} f_{XY}(u, v)dudv$$

If both are jointly discrete then we have the *joint probability mass function* defined as

$$p(x, y) \equiv P[X = x, Y = y] \qquad (1.36)$$

and the corresponding marginal mass functions can be computed as

$$P[X = x] = \sum_{y} p(x, y)$$

$$P[Y = y] = \sum_{x} p(x, y) \qquad (1.37)$$

With the definitions of joint distribution and density function in place, we are now in a position to extend the notion of statistical independence to two random variables. Basically, two random variables X and Y are said to be statistically independent if the events $\{x \in E\}$ and $\{y \in F\}$ are independent, i.e.:

$$P[x \in E, y \in F] = P[x \in E] \cdot P[y \in F]$$

From the above expression, it can be deduced that X and Y are statistically independent if any of the following relationships hold:

- $F_{XY}(x, y) = F_X(x) \cdot F_Y(y)$
- $f_{XY}(x, y) = f_X(x) \cdot f_Y(y)$ if both are jointly continuous
- $P[x = x, Y = y] = P[X = x] \cdot P[Y = y]$ if both are jointly discrete

We summarize below some of the properties pertaining to the relationships between two random variables. In the following, X and Y are two independent random variables defined on the same sample space, c is a constant and g and h are two arbitrary real functions.

(i)   Convolution Property
      If $Z = X + Y$, then

      • if X and Y are jointly discrete

$$P[Z = k] = \sum_{i+j=k} P[X = i]P[Y = j] = \sum_{i=0}^{k} P[X = i]P[Y = k - i] \qquad (1.38)$$

- if X and Y are jointly continuous

$$f_Z(z) = \int_0^\infty f_X(x)f_Y(z-x)dx = \int_0^\infty f_X(z-y)f_Y(y)dy$$
$$= f_X(x) \otimes f_Y(y) \qquad (1.39)$$

where $\otimes$ denotes the convolution operator.

(ii) $E[cX] = cE[X]$
(iii) $E[X + Y] = E[X] + E[Y]$
(iv) $E[g(X)h(Y)] = E[g(X)] \cdot E[h(Y)]$ if X and Y are independent
(v) $Var[cX] = c^2Var[X]$
(vi) $Var[X + Y] = Var[X] + Var[Y]$ if X and Y are independent
(vi) $Var[X] = E[X^2] - (E[X])^2$

Example 1.7: Random sum of random variables

Consider the voice packetization process during a teleconferencing session, where voice signals are packetized at a teleconferencing station before being transmitted to the other party over a communication network in packet form. If the number (N) of voice signals generated during a session is a random variable with mean $E(N)$, and a voice signal can be digitized into X packets, find the mean and variance of the number of packets generated during a teleconferencing session, assuming that these voice signals are identically distributed.

Solution

Denote the number of packets for each voice signal as X_i and the total number of packets generated during a session as T, then we have

$$T = X_1 + X_2 + \ldots + X_N$$

To calculate the expected value, we first condition it on the fact that $N = k$ and then use the total probability theorem to sum up the probability. That is:

$$E[T] = \sum_{i=1}^N E[T|N = k]P[N = k]$$
$$= \sum_{i=1}^N kE[X]P[N = k]$$
$$= E[X]E[N]$$

To compute the variance of T, we first compute $E[T^2]$:

$$E[T^2|N = k] = Var[T|N = k] + (E[T|N = k])^2$$
$$= kVar[X] + k^2(E[X])^2$$

and hence we can obtain

$$E[T^2] = \sum_{k=1}^{N}(kVar[X] + k^2(E[X])^2)P[N = k]$$
$$= Var[X]E[N] + E[N^2](E[X])^2$$

Finally:

$$Var[T] = E[T^2] - (E[T])^2$$
$$= Var[X]E[N] + E[N^2](E[X])^2 - (E[X])^2(E[N])^2$$
$$= Var[X]E[N] + (E[X])^2Var[N]$$

Example 1.8

Consider two packet arrival streams to a switching node, one from a voice source and the other from a data source. Let X be the number of time slots until a voice packet arrives and Y the number of time slots till a data packet arrives. If X and Y are geometrically distributed with parameters p and q respectively, find the distribution of the time (in terms of time slots) until a packet arrives at the node.

Solution

Let Z be the time until a packet arrives at the node, then $Z = \min(X, Y)$ and we have

$$P[Z > k] = P[X > k]P[Y > k]$$
$$1 - F_Z(k) = \{1 - F_X(k)\}\{1 - F_Y(k)\}$$

but

$$F_X(k) = \sum_{j=1}^{\infty}p(1-p)^{j-1} = p\frac{1-(1-p)^k}{1-(1-p)}$$
$$= 1-(1-p)^k$$

Similarly

$$F_Y(k) = 1 - (1 - q)^k$$

Therefore, we obtain

$$F_Z(k) = 1 - (1-p)^k(1-q)^k$$
$$= 1 - [(1-p)(1-q)]^k$$

Theorem 1.1

Suppose a random variable Y is a function of a finite number of independent random variables $\{X_i\}$, with arbitrary known probability density functions (*pdf*). If

$$Y = \sum_{i=1}^{n} X_i$$

then the *pdf* of Y is given by the density function:

$$g_Y(y) = f_{X1}(x_1) \otimes f_{X2}(x_2) \otimes f_{X3}(x_3) \otimes \dots f_{Xn}(x_n) \qquad (1.40)$$

The keen observer might note that this result is a general extension of expression (1.39). Fortunately the convolution of density functions can be easily handled by transforms (z or Laplace).

Example 1.9

Suppose the propagation delay along a link follows the exponential distribution:

$$f_X(x_i) = \exp(-x_i) \quad \text{for } x_i \geq 0 \quad \text{for } i = 1 \dots 10.$$

Find the density function $g(y)$ where $y = x_1 + x_2 + \dots x_{10}$.

Solution

Consider the effect of the new random variable by using Theorem 1.1 above, where each exponential random variable are independent and identically distributed with $g(y) = \dfrac{y^{i-1}e^{-y}}{(i-1)!}$ for $y \geq 0$ as shown in Figure 1.5.

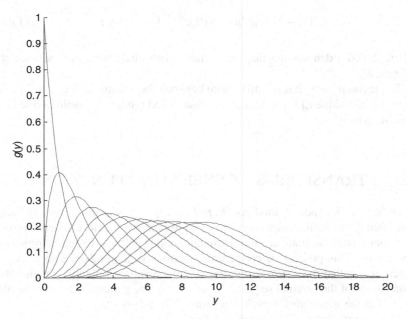

Figure 1.5   The density function $g(y)$ for $I = 1 \ldots 10$

1.1.6   Independence of Random Variables

Independence is probably the most fertile concept in probability theorems, for example, it is applied to queueing theory under the guise of the well-known Kleinrock independence assumption.

Theorem 1.2

[Strong law of large numbers]
For n independent and identically distributed random variables $\{X_n, n \geq 1\}$:

$$Y_n = \{X_1 + X_2 \ldots X_n\}/n \to E[X_1] \quad \text{as} \quad n \to \infty \qquad (1.41)$$

That is, for large n, the arithmetic mean of Y_n of n independent and identically distributed random variables with the same distribution is close to the expected value of these random variables.

Theorem 1.3

[Central Limit theorem]
Given Y_n as defined above:

$$\{Y_n - E[X_1]\}\sqrt{n} \approx N(0, \sigma^2) \quad \text{for} \quad n \gg 1 \tag{1.42}$$

where $N(0, \sigma^2)$ denotes the random variable with mean zero and variance σ^2 of each X_n.

The theorem says that the difference between the arithmetic mean of Y_n and the expected value $E[X_1]$ is a Gaussian distributed random variable divided by $\sqrt{n}$ for large n.

1.2 z-TRANSFORMS – GENERATING FUNCTIONS

If we have a sequence of numbers $\{f_0, f_1, f_2, \ldots f_k \ldots\}$, possibly infinitely long, it is often desirable to compress it into a single function – a closed-form expression that would facilitate arithmetic manipulations and mathematical proofing operations. This process of converting a sequence of numbers into a single function is called the *z-transformation*, and the resultant function is called the *z*-transform of the original sequence of numbers. The *z*-transform is commonly known as the *generating function* in probability theory.

The *z*-transform of a sequence is defined as

$$F(z) \equiv \sum_{k=0}^{\infty} f_k z^k \tag{1.43}$$

where z^k can be considered as a 'tag' on each term in the sequence and hence its position in that sequence is uniquely identified should the sequence need to be recovered. The *z*-transform $F(z)$ of a sequence exists so long as the sequence grows slower than a^k, i.e.:

$$\lim_{k \to \infty} \frac{|k_k|}{a^k} = 0$$

for some $a > 0$ and it is unique for that sequence of numbers.

z-transform is very useful in solving difference equations (or so-called recursive equations) that contain constant coefficients. A difference equation is an equation in which a term (say kth) of a function $f(\bullet)$ is expressed in terms of other terms of that function. For example:

$$f_{k-1} + f_{k+1} = 2f_k$$

This kind of difference equation occurs frequently in the treatment of queueing systems. In this book, we use $\Leftrightarrow$ to indicate a transform pair, for example, $f_k \Leftrightarrow F(z)$.

1.2.1 Properties of z-Transforms

z-transform possesses some interesting properties which greatly facilitate the evaluation of parameters of a random variable. If X and Y are two independent random variables with respective probability mass functions f_k and f_g, and their corresponding transforms $F(z)$ and $G(z)$ exist, then we have the two following properties:

(a) Linearity property

$$af_k + bg_k \Leftrightarrow aF(z) + bG(z) \qquad (1.44)$$

This follows directly from the definition of z-transform, which is a linear operation.

(b) Convolution property
If we define another random variable $H = X + Y$ with a probability mass function h_k, then the z-transform $H(z)$ of h_k is given by

$$H(z) = F(z) \cdot G(z) \qquad (1.45)$$

The expression can be proved as follows:

$$H(z) = \sum_{k=0}^{\infty} h_k z^k$$
$$= \sum_{k=0}^{\infty} \sum_{i=0}^{k} f_i g_{k-i} z^k$$
$$= \sum_{i=0}^{\infty} \sum_{k=i}^{\infty} f_i g_{k-i} z^k$$
$$= \sum_{i=0}^{\infty} f_i z^i \sum_{k=i}^{\infty} g_{k-i} z^{k-i}$$
$$= F(z) \cdot G(z)$$

The interchange of summary signs can be viewed from the following:

		Index i		
Index	$f_0 g_0$			
k	$f_0 g_1$	$f_1 g_0$		
$\downarrow$	$f_0 g_2$	$f_1 g_1$	$\cdot$	
	$\cdots$	$\cdots$	$\cdots$	$\cdot$

Table 1.2   Some z-transform pairs

Sequence	z-transform
$u_k = 1 \; k = 0, 1, 2 \ldots$	$1/(1-z)$
u_{k-a}	$z^a/(1-z)$
Aa^k	$A/(1-az)$
ka^k	$az/(1-az)^2$
$(k+1)a^k$	$1/(1-az)^2$
$a/k!$	ae^z

(c) Final values and expectation

(i)
$$F(z)|_{z=1} = 1 \tag{1.46}$$

(ii)
$$E[X] = \frac{d}{dz} F(z)\Big|_{z=1} \tag{1.47}$$

(iii)
$$E[X^2] = \frac{d^2}{dz^2} F(z)\Big|_{z=1} + \frac{d}{dz} F(z)\Big|_{z=1} \tag{1.48}$$

Table 1.2 summarizes some of the z-transform pairs that are useful in our subsequent treatments of queueing theory.

Example 1.10

Let us find the z-transforms for Binomial, Geometric and Poisson distributions and then calculate the expected values, second moments and variances for these distributions.

(i)   Binomial distribution:

$$B_X(z) = \sum_{k=0}^{n} \binom{n}{k} p^k (1-p)^{n-k} z^k$$
$$= (1 - p + pz)^n$$

$$\frac{d}{dz} B_X(z) = np(1 - p + pz)^{n-1}$$

therefore

$$E[X] = \frac{d}{dz} B_X(z)\Big|_{z=1} = np$$

and

$$\frac{d^2}{dz^2}B_X(z) = np(n-1)p(1-p+pz)^{n-2}$$

$$E[X^2] = n(n-1)p^2 + np$$

$$\sigma^2 = E[X^2] - E^2[X]$$
$$= np(1-p)$$

(ii) Geometric distribution:

$$G(z) = \sum_{k=1}^{\infty} p(1-p)^{k-1}z^k = \frac{pz}{1-(1-p)z}$$

$$E[X] = \frac{p}{1-(1-p)z} + \frac{pz(1-p)}{(1-(1-p)z)^2}\bigg|_{z=1} = \frac{1}{p}$$

$$\frac{d^2}{dz^2}G(z)\bigg|_{z=1} = 2\left(\frac{1}{p^2} - \frac{1}{p}\right)$$

$$\sigma^2 = \frac{1}{p^2} - \frac{1}{p}$$

(iii) Poisson distribution:

$$G(z) = \sum_{k=0}^{\infty} \frac{(\lambda t)^k}{k!}e^{-\lambda t}z^k = e^{-\lambda t}e^{+\lambda tz} = e^{-\lambda t(1-z)}$$

$$E[X] = \frac{d}{dz}G(z)\bigg|_{z=1} = \lambda te^{-\lambda t(1-z)} = \lambda t$$

$$\frac{d^2}{dz^2}G(z)\bigg|_{z=1} = (\lambda t)^2$$

$$\sigma^2 = E[X]^2 - E^2[X] = \lambda t$$

Table 1.3 summarizes the z-transform expressions for those probability mass functions discussed in Section 1.2.3.

Table 1.3   z-transforms for some of the discrete random variables

Random variable	z-transform
Bernoulli	$G(z) = q + pz$
Binomial	$G(z) = (q + pz)^n$
Geometric	$G(z) = pz/(1 - qz)$
Poisson	$G(z) = e^{-\lambda t(1-z)}$

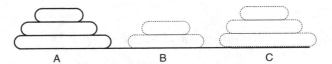

Figure 1.6   A famous legendary puzzle

Example 1.11

This is a famous legendary puzzle. According to the legend, a routine morning exercise for Shaolin monks is to move a pile of iron rings from one corner (Point A) of the courtyard to another (Point C) using only a intermediate point (Point B) as a resting point (Figure 1.6). During the move, a larger ring cannot be placed on top of a smaller one at the resting point. Determine the number of moves required if there are k rings in the pile.

Solution

To calculate the number of moves (m_k) required, we first move the top $(k-1)$ rings to Point B and then move the last ring to Point C, and finally move the $(k-1)$ rings from Point B to Point C to complete the exercise. Denote its z-transform as $M(z) = \sum_{k=0}^{\infty} m_k z^k$ and $m_0 = 0$, then from the above-mentioned recursive approach we have

$$m_k = m_{k-1} + 1 + m_{k-1} \qquad k \geq 1$$
$$m_k = 2m_{k-1} + 1$$

Multiplying the equation by z^k and summing it from zero to infinity, we have

$$\sum_{k=1}^{\infty} m_k z^k = 2 \sum_{k=1}^{\infty} m_{k-1} z^k + \sum_{k=1}^{\infty} z^k$$
$$M(z) - m_0 = 2zM(z) + \frac{z}{1-z}$$

and

$$M(z) = \frac{z}{(1-z)(1-2z)}$$

To find the inverse of this expression, we do a partial fraction expansion:

$$M(z) = \frac{1}{1-2z} + \frac{-1}{1-z} = (2-1)z + (2^2-1)z^2 + (2^3-1)z^3 + \ldots$$

Therefore, we have $m_k = 2^k - 1$

Example 1.12

Another well-known puzzle is the Fibonacci numbers {1, 1, 2, 3, 5, 8, 13, 21, …}, which occur frequently in studies of population grow. This sequence of numbers is defined by the following recursive equation, with the initial two numbers as $f_0 = f_1 = 1$:

$$f_k = f_{k-1} + f_{k-2} \quad k \geq 2$$

Find an explicit expression for f_k.

Solution

First multiply the above equations by z^k and sum it to infinity, so we have

$$\sum_{k=2}^{\infty} f_k z^k = \sum_{k=2}^{\infty} f_{k-1} z^k + \sum_{k=2}^{\infty} f_{k-2} z^k$$

$$F(z) - f_1 z - f_0 = z(F(z) - f_0) + z^2 F(z)$$

$$F(z) = \frac{-1}{z^2 + z - 1}$$

Again, by doing a partial fraction expression, we have

$$F(z) = \frac{1}{\sqrt{5}\, z_1 [1 - (z/z_1)]} - \frac{1}{\sqrt{5}\, z_2 [1 - (z/z_2)]}$$

$$= \frac{1}{\sqrt{5}\, z_1} \left(1 + \frac{z}{z_1} + \ldots \right) - \frac{1}{\sqrt{5}\, z_2} \left(1 + \frac{z}{z_2} + \ldots \right)$$

where

$$z_1 = \frac{-1+\sqrt{5}}{2} \quad \text{and} \quad z_2 = \frac{-1-\sqrt{5}}{2}$$

Therefore, picking up the k term, we have

$$f_k = \frac{1}{\sqrt{5}}\left[\left(\frac{1+\sqrt{5}}{2}\right)^{k+1} - \left(\frac{1-\sqrt{5}}{2}\right)^{k+1}\right]$$

1.3   LAPLACE TRANSFORMS

Similar to z-transform, a continuous function $f(t)$ can be transformed into a new complex function to facilitate arithmetic manipulations. This transformation operation is called the Laplace transformation, named after the great French mathematician Pierre Simon Marquis De Laplace, and is defined as

$$F(s) = L[f(t)] = \int_{-\infty}^{\infty} f(t)e^{-st}dt \qquad (1.49)$$

where s is a complex variable with real part σ and imaginary part $j\omega$, i.e. $s = \sigma + j\omega$ and $j = \sqrt{-1}$. In the context of probability theory, all the density functions are defined only for the real-time axis, hence the 'two-sided' Laplace transform can be written as

$$F(s) = L[f(t)] = \int_{0^-}^{\infty} f(t)e^{-st}dt \qquad (1.50)$$

with the lower limit of the integration written as 0^- to include any discontinuity at $t = 0$. This Laplace transform will exist so long as $f(t)$ grows no faster than an exponential, i.e.:

$$|f(t)| \le Me^{\alpha t}$$

for all $t \ge 0$ and some positive constants M and α. The original function $f(t)$ is called the inverse transform or inverse of $F(s)$, and is written as

$$f(t) = L^{-1}[F(s)]$$

The Laplace transformation is particularly useful in solving differential equations and corresponding initial value problems. In the context of queueing theory, it provides us with an easy way of finding performance measures of a queueing system in terms of Laplace transforms. However, students should note that it is at times extremely difficult, if not impossible, to invert these Laplace transform expressions.

1.3.1 Properties of the Laplace Transform

The Laplace transform enjoys many of the same properties as the z-transform as applied to probability theory. If X and Y are two independent continuous random variables with density functions $f_X(x)$ and $f_Y(y)$, respectively and their corresponding Laplace transforms exist, then their properties are:

(i) Uniqueness property

$$f_X(\tau) = f_Y(\tau) \quad \text{implies} \quad F_X(s) = F_Y(s) \tag{1.51}$$

(ii) Linearity property

$$af_X(x) + bf_Y(y) \Rightarrow aF_X(s) + bF_Y(s) \tag{1.52}$$

(iii) Convolution property

If $Z = X + Y$, then

$$F_Z(s) = L[f_z(z)] = L[f_{X+Y}(x+y)]$$
$$= F_X(s) \cdot F_Y(s) \tag{1.53}$$

(iv) Expectation property

$$E[X] = -\frac{d}{ds} F_X(s) \Big|_{s=0} \quad \text{and} \quad E[X^2] = \frac{d^2}{ds^2} F_X(s) \Big|_{s=0} \tag{1.54}$$

$$E[X^n] = (-1)^n \frac{d^n}{ds^n} F_X(s) \Big|_{s=0} \tag{1.55}$$

(v) Differentiation property

$$L[f_X'(x)] = sF_X(s) - f_X(0) \tag{1.56}$$

$$L[f_X''(x)] = s^2 F_X(s) - sf_X(0) - f_X'(0) \tag{1.57}$$

Table 1.4 shows some of the Laplace transform pairs which are useful in our subsequent discussions on queueing theory.

Table 1.4   Some Laplace transform pairs

Function	Laplace transform
$\delta(t)$ *unit impulse*	1
$\delta(t - a)$	e^{-as}
1 *unit step*	$1/s$
t	$1/s^2$
$t^{n-1}/(n-1)!$	$1/s^n$
Ae^{at}	$A/(s-a)$
te^{at}	$1/(s-a)^2$
$t^{n-1}e^{at}/(n-1)!$	$1/(s-a)^n \; n = 1,2,\ldots$

Example 1.13

Derive the Laplace transforms for the exponential and k-stage Erlang probability density functions, and then calculate their means and variances.

(i) exponential distribution

$$F(s) = \int_0^\infty e^{-ss} \lambda e^{-\lambda x} dx = \left[-\frac{\lambda}{s+\lambda} e^{-(s+\lambda)x} \right]_0^\infty = \frac{\lambda}{s+\lambda}$$

$$F(s) = \int_0^\infty e^{-sx} \lambda e^{-\lambda x} dx = \left[-\frac{\lambda}{s+\lambda} e^{-(s+\lambda)x} \right]_0^\infty = \frac{\lambda}{s+\lambda}$$

$$E[X] = -\frac{d}{ds} F(s) \bigg|_{s=0} = \frac{1}{\lambda}$$

$$E[X^2] = \frac{d^2}{ds^2} F(S) \bigg|_{s=0} = \frac{2}{\lambda^2}$$

$$\sigma^2 = E[X^2] - E^2[X] = \frac{1}{\lambda^2}$$

$$C = \frac{\sigma}{E[X]} = 1$$

(ii) k-stage Erlang distribution

$$F(s) = \int_0^\infty e^{-ss} \frac{\lambda^k x^{k-1}}{(k-1)!} e^{-\lambda x} dx = \frac{\lambda^k}{(k-1)!} \int_0^\infty x^{k-1} e^{-(s+\lambda)x} dx$$

$$= \frac{\lambda^k}{(s+\lambda)^k (k-1)!} \int_0^\infty \{(s+\lambda)x\}^{k-1} e^{-(s+\lambda)x} d(s+\lambda)x$$

Table 1.5 Laplace transforms for some probability functions

Random variable	Laplace transform
Uniform $a < x < b$	$F(s) = e^{-s(a+b)}/s(b - a)$
Exponent	$F(s) = \lambda/s + \lambda$
Gamma	$F(s) = \lambda^\alpha/(s + \lambda)^\alpha$
Erlang-k	$F(s) = \lambda^k/(s + \lambda)^k$

The last integration term is recognized as the gamma function and is equal to $(k - 1)!$ Hence we have

$$F(s) = \left(\frac{\lambda}{s + \lambda}\right)^k$$

Table 1.5 gives the Laplace transforms for those continuous random variables discussed in Section 1.1.3.

Example 1.14

Consider a counting process whose behavior is governed by the following two differential-difference equations:

$$\frac{d}{dt}P_k(t) = -\lambda P_k(t) + \lambda P_{k-1}(t) \qquad k > 0$$

$$\frac{d}{dt}P_0(t) = -\lambda P_0(t)$$

Where $P_k(t)$ is the probability of having k arrivals within a time interval $(0, t)$ and λ is a constant, show that $P_k(t)$ is Poisson distributed.

Let us define the Laplace transform of $P_k(t)$ and $P_0(t)$ as

$$F_k(s) = \int_0^\infty e^{-st} P_k(t) dt$$

$$F_0(s) = \int_0^\infty e^{-st} P_0(t) dt$$

From the properties of Laplace Transform, we know

$$L[P_k'(t)] = sF_k(s) - P_k(0)$$

$$L[P_0'(t)] = sF_0(s) - P_0(0)$$

Substituting them into the differential-difference equations, we obtain

$$F_0(s) = \frac{P_0(0)}{s + \lambda}$$

$$F_k(s) = \frac{P_k(0) + \lambda L_{k-1}(s)}{s + \lambda}$$

If we assume that the arrival process begins at time $t = 0$, then $P_0(0) = 1$ and $P_k(0) = 0$, and we have

$$F_0(s) = \frac{1}{s + \lambda}$$

$$F_k(s) = \frac{\lambda}{s + \lambda} F_{k-1}(s) = \left(\frac{\lambda}{s + \lambda}\right)^k F_0(s)$$

$$= \frac{\lambda^k}{(s + \lambda)^{k+1}}$$

Inverting the two transforms, we obtain the probability mass functions:

$$P_0(0) = e^{-\lambda t}$$

$$P_k(t) = \frac{(\lambda t)^k}{k!} e^{-\lambda t}$$

1.4   MATRIX OPERATIONS

In Chapter 8, with the introduction of Markov-modulated arrival models, we will be moving away from the familiar Laplace (z-transform) solutions to a new approach of solving queueing systems, called matrix-geometric solutions. This particular approach to solving queueing systems was pioneered by Marcel F Neuts. It takes advantage of the similar structure presented in many interesting stochastic models and formulates their solutions in terms of the solution of a nonlinear matrix equation.

1.4.1   Matrix Basics

A matrix is a $m \times n$ rectangular array of real (or complex) numbers enclosed in parentheses, as shown below:

$$\tilde{A} = (a_{ij}) = \begin{pmatrix} a_{11} & a_{12} & \cdots & a_{1n} \\ a_{21} & a_{22} & \cdots & a_{2n} \\ \vdots & & & \vdots \\ a_{m1} & a_{m2} & \cdots & a_{mn} \end{pmatrix}$$

where a_{ij}'s are the elements (or components) of the matrix. A $m \times 1$ matrix is a column vector and a $1 \times n$ matrix is a row vector. In the sequel, we denote matrices by capital letters with a tilde (~) on top, such as $\tilde{A}, \tilde{B}$ & $\tilde{C}$, column vectors by small letters with a tilde, such as $\tilde{f}$ & $\tilde{g}$, and row vectors by small Greek letters, such as $\tilde{\pi}$ & $\tilde{\nu}$.

A matrix whose elements are all zero is called the null matrix and denoted by $\tilde{0}$. A diagonal matrix $(\tilde{\Lambda})$ is a square matrix whose entries other than those in the diagonal positions are all zero, as shown below:

$$\tilde{\Lambda} = diag(a_{11}, a_{12}, \ldots, a_{nn})$$

$$= \begin{pmatrix} a_{11} & 0 & \cdots & 0 \\ 0 & a_{12} & \cdots & 0 \\ \cdots & & \ddots & \\ 0 & 0 & \cdots & a_{nn} \end{pmatrix}$$

If the diagonal entries are all equal to one then we have the identity matrix $(\tilde{I})$.

The transpose $\tilde{A}^T$ of a $m \times n$ matrix $\tilde{A} = (a_{ij})$ is the $n \times m$ matrix obtained by interchanging the rows and columns of $\tilde{A}$, that is

$$\tilde{A}^T = (a_{ji}) = \begin{pmatrix} a_{11} & a_{21} & \cdots & a_{m1} \\ a_{12} & a_{22} & \cdots & a_{m2} \\ \cdots & & & \vdots \\ a_{1n} & a_{2n} & \cdots & a_{mn} \end{pmatrix}$$

The inverse of an n-rowed square matrix $\tilde{A}$ is denoted by $\tilde{A}^{-1}$ and is an n-rowed square matrix that satisfies the following expression:

$$\tilde{A}\tilde{A}^{-1} = \tilde{A}^{-1}\tilde{A} = \tilde{I}$$

$\tilde{A}^{-1}$ exists (and is then unique) if and only if A is non-singular, i.e. if and only if the determinant of A is not zero, $\tilde{A} \neq 0$. In general, the inverse of $\tilde{A}$ is given by

$$\tilde{A}^{-1} = \frac{1}{\det \tilde{A}} \begin{pmatrix} A_{11} & A_{21} & \cdots & A_{n1} \\ A_{12} & A_{22} & & \\ \cdots & & & \cdots \\ A_{1n} & A_{2n} & & A_{nn} \end{pmatrix}$$

where A_{ij} is the *cofactor* of a_{ij} in $\tilde{A}$. The cofactor of a_{ij} is the product of $(-1)^{i+j}$ and the determinant formed by deleting the ith row and the jth column from the det $\tilde{A}$. For a 2×2 matrix $\tilde{A}$, the inverse is given by

$$\tilde{A} = \begin{pmatrix} a_{11} & a_{12} \\ a_{21} & a_{22} \end{pmatrix} \quad \text{and} \quad \tilde{A}^{-1} = \frac{1}{a_{11}a_{22} - a_{21}a_{12}} \begin{pmatrix} a_{22} & -a_{12} \\ -a_{21} & a_{11} \end{pmatrix}$$

We summarize some of the properties of matrixes that are useful in manipulating them. In the following expressions, α and β are numbers:

(i) $\alpha(\tilde{A} + \tilde{B}) = \alpha\tilde{A} + \alpha\tilde{B}$ and $(\alpha + \beta)\tilde{A} = \alpha\tilde{A} + \beta\tilde{A}$

(ii) $(\alpha\tilde{A})\tilde{B} = \alpha(\tilde{A}\tilde{B}) = \tilde{A}(\alpha\tilde{B})$ and $\tilde{A}(\tilde{B}\tilde{C}) = (\tilde{A}\tilde{B})\tilde{C}$

(iii) $(\tilde{A} + \tilde{B})\tilde{C} = \tilde{A}\tilde{C} + \tilde{B}\tilde{C}$ and $\tilde{C}(\tilde{A} + \tilde{B}) = \tilde{C}\tilde{A} + \tilde{C}\tilde{B}$

(iv) $\tilde{A}\tilde{B} \neq \tilde{B}\tilde{A}$ in general

(v) $\tilde{A}\tilde{B} = \tilde{0}$ does not necessarily imply $\tilde{A} = \tilde{0}$ *or* $\tilde{B} = \tilde{0}$

(vi) $(\tilde{A} + \tilde{B})^T = \tilde{A}^T + \tilde{B}^T$ and $(\tilde{A}^T)^T = \tilde{A}$

(vii) $(\tilde{A}\tilde{B})^T = \tilde{B}^T\tilde{A}^T$ and $\det\tilde{A} = det\tilde{A}^T$

(viii) $(\tilde{A}^{-1})^{-1} = \tilde{A}$ and $(\tilde{A}\tilde{B})^{-1} = \tilde{B}^{-1}\tilde{A}^{-1}$

(ix) $(\tilde{A}^{-1})^T = (\tilde{A}^T)^{-1}$ and $(\tilde{A}^2)^{-1} = (\tilde{A}^{-1})^2$

1.4.2   Eigenvalues, Eigenvectors and Spectral Representation

An eigenvalue (or characteristic value) of an $n \times n$ square matrix $\tilde{A} = (a_{ij})$ is a real or complex scalar λ satisfying the following vector equation for some non-zero (column) vector $\tilde{x}$ of dimension $n \times 1$. The vector $\tilde{x}$ is known as the eigenvector, or more specifically the column (or right) eigenvector:

$$\tilde{A}\tilde{x} = \lambda\tilde{x} \tag{1.58}$$

This equation can be rewritten as $(\tilde{A} - \lambda\tilde{I})\tilde{x} = 0$ and has a non-zero solution $\tilde{x}$ only if $(\tilde{A} - \lambda\tilde{I})$ is singular; that is to say that any eigenvalue must satisfy $\det(\tilde{A} - \lambda\tilde{I}) = 0$. This equation, $\det(\tilde{A} - \lambda\tilde{I}) = 0$, is a polynomial of degree n in λ and has exactly n real or complex roots, including multiplicity. Therefore,

A has n eigenvalues λ_1, λ_2, ..., λ_n with the corresponding eigenvectors $\tilde{x}_1$, $\tilde{x}_2$, ..., $\tilde{x}_n$. The polynomial is known as the characteristic polynomial of A and the set of eigenvalues is called the spectrum of $\tilde{A}$.

Similarly, the row (or left) eigenvectors are the solutions of the following vector equation:

$$\tilde{\pi}\tilde{A} = \lambda\tilde{\pi} \tag{1.59}$$

and everything that is said about column eigenvectors is also true for row eigenvectors.

Here, we summarize some of the properties of eigenvalues and eigenvectors:

(i)   The sum of the eigenvalues of $\tilde{A}$ is equal to the sum of the diagonal entries of $\tilde{A}$. The sum of the diagonal entries of $\tilde{A}$ is called the trace of $\tilde{A}$.

$$tr(\tilde{A}) = \sum_i \lambda_i \tag{1.60}$$

(ii) If A has eigenvalues λ_1, λ_2, ..., λ_n, then λ_1^k, λ_2^k, ..., λ_n^k are eigenvectors of $\tilde{A}^k$, and we have

$$tr(\tilde{A}^k) = \sum_i \lambda_i^k \quad k = 1, 2, \ldots \tag{1.61}$$

(iii) If $\tilde{A}$ is a non-singular matrix with eigenvalues λ_1, λ_2, ..., λ_n, then $\lambda_1^{-1}>$, λ_2^{-1}), ..., λ_n^{-1} are eigenvectors of $\tilde{A}^{-1}$. Moreover, any eigenvector of $\tilde{A}$ is an eigenvector of $\tilde{A}^{-1}$.

(iv) $\tilde{A}$ and $\tilde{A}^T$ do not necessarily have the same eigenvectors. However, if $\tilde{A}^T\tilde{x} = \lambda\tilde{x}$ then $\tilde{x}^T\tilde{A} = \lambda\tilde{x}^T$, and the row vector $\tilde{x}^T$ is called a left eigenvector of $\tilde{A}$.

It should be pointed out that eigenvalues are in general relatively difficult to compute, except for certain special cases.

If the eigenvalues λ_1, λ_2, ..., λ_n of a matrix $\tilde{A}$ are all distinct, then the corresponding eigenvectors $\tilde{x}_1$, $\tilde{x}_2$, ..., $\tilde{x}_n$ are linearly independent, and we can express $\tilde{A}$ as

$$\tilde{A} = \tilde{N}\tilde{\Lambda}\tilde{N}^{-1} \tag{1.62}$$

where $\tilde{\Lambda} = diag(\lambda_1, \lambda_2, \ldots, \lambda_n)$, $\tilde{N} = [\tilde{x}_1, \tilde{x}_2, \ldots, \tilde{x}_n]$ whose ith column is $\tilde{x}_i$, $\tilde{N}^{-1}$ is the inverse of $\tilde{N}$, and is given by

$$\tilde{N}^{-1} = \begin{bmatrix} \tilde{\pi}_1 \\ \tilde{\pi}_2 \\ \dots \\ \tilde{\pi}_n \end{bmatrix}$$

By induction, it can be shown that $\tilde{A}^k = \tilde{N}\tilde{\Lambda}^k\tilde{N}^{-1}$.

If we define $\tilde{B}_k$ to be the matrix obtained by multiplying the column vector $\tilde{x}_k$ with the row vector $\tilde{\pi}_k$, then we have

$$\begin{aligned}\tilde{B}_k &= \tilde{x}_k\tilde{\pi}_k \\ &= \begin{pmatrix} x_k(1)\pi_k(1) & \dots & x_k(1)\pi_k(n) \\ \dots & & \dots \\ x_k(n)\pi_k(1) & \dots & x_k(n)\pi_k(n) \end{pmatrix}\end{aligned} \tag{1.63}$$

It can be shown that

$$\begin{aligned}\tilde{A} &= \tilde{N}\tilde{\Lambda}\tilde{N}^{-1} \\ &= \lambda_1\tilde{B}_1 + \lambda_2\tilde{B}_2 + \dots + \lambda_n\tilde{B}_n\end{aligned} \tag{1.64}$$

and

$$\tilde{A}^k = \lambda_{11}^k\tilde{B}_1 + \lambda_2^k\tilde{B}_2 + \dots + \lambda_n^k\tilde{B}_n \tag{1.65}$$

The expression of $\tilde{A}$ in terms of its eigenvalues and the matrices $\tilde{B}_k$ is called the spectral representation of $\tilde{A}$.

1.4.3 Matrix Calculus

Let us consider the following set of ordinary differential equations with constant coefficients and given initial conditions:

$$\frac{d}{dt}x_1(t) = a_{11}x_1(t) + a_{11}x_2(t) + \dots + a_{1n}x_n(t)$$
$$\vdots$$
$$\frac{d}{dt}x_n(t) = a_{n1}x_1(t) + a_{n2}x_2(t) + \dots + a_{nn}x_n(t) \tag{1.66}$$

In matrix notation, we have

$$\tilde{x}(t)' = \tilde{A}\tilde{x}(t) \tag{1.67}$$

where $\tilde{x}(t)$ is a $n \times 1$ vector whose components $x_i(t)$ are functions of an independent variable t, and $x(t)'$ denotes the vector whose components are the derivatives dx_i/dt. There are two ways of solving this vector equation:

(i) First let us assume that $\tilde{x}(t) = e^{\lambda t}\tilde{p}$, where $\tilde{P}$ is a scalar vector and substitute it in Equation (1.67), then we have

$$\lambda e^{\lambda t}\tilde{p} = \tilde{A}(e^{\lambda t}\tilde{p})$$

Since $e^{\lambda t} \neq 0$, it follows that λ and $\tilde{p}$ must satisfy $\tilde{A}\tilde{p} = \tilde{\lambda}\tilde{p}$; therefore, if λ_i is an eigenvector of A and $\tilde{p}_i$ is a corresponding eigenvector, then $e^{\lambda_i t}\tilde{p}_i$ is a solution. The general solution is given by

$$\tilde{x}(t) = \sum_{i=1}^{n}\alpha_i e^{\lambda_i t}\tilde{p}_i \tag{1.68}$$

where α_i is the constant chosen to satisfy the initial condition of Equation (1.67).

(ii) The second method is to define the matrix exponential $e^{\tilde{A}t}$ through the convergent power series as

$$\exp\{\tilde{A}t\} = \sum_{k=0}^{\infty}\frac{(\tilde{A}t)^k}{k!} = I + \tilde{A}t + \frac{(\tilde{A}t)^2}{2!} + \ldots + \frac{(\tilde{A}t)^k}{k!} \tag{1.69}$$

By differentiating the expression with respect to t directly, we have

$$\frac{d}{dt}(e^{\tilde{A}t}) = \tilde{A} + \tilde{A}^2 t + \frac{\tilde{A}^3 t^2}{2!} + \ldots$$

$$= \tilde{A}\left(\tilde{I} + \tilde{A}t + \frac{\tilde{A}^2 t^2}{2!} + \ldots\right) = \tilde{A}e^{\tilde{A}t}$$

Therefore, $e^{\tilde{A}t}$ is a solution to Equation (1.67) and is called the fundamental matrix for (1.67).

We summarize some of the useful properties of the matrix exponential below:

(i) $e^{\tilde{A}(s+t)} = e^{\tilde{A}s}e^{\tilde{A}t}$

(ii) $e^{\tilde{A}t}$ is never singular and its inverse is $e^{-\tilde{A}t}$

(iii) $e^{(\tilde{A}+\tilde{B})t} = e^{\tilde{A}t}e^{\tilde{B}t}$ for all t, only if $\tilde{A}\tilde{B} = \tilde{B}\tilde{A}$

(iv) $\dfrac{d}{dt}e^{\tilde{A}t} = \tilde{A}e^{\tilde{A}t} = e^{\tilde{A}t}\tilde{A}$

(v) $(\tilde{I} - \tilde{A})^{-1} = \sum\limits_{i=0}^{\infty}\tilde{A}^i = \tilde{I} + \tilde{A} + \tilde{A}^2 + \ldots$

(vi) $e^{\tilde{A}t} = \tilde{N}e^{\tilde{\Lambda}t}\tilde{N}^{-1}$
$\qquad = e^{\lambda_1 t}\tilde{B}_1 + e^{\lambda_2 t}\tilde{B}_2 + \ldots + e^{\lambda_n t}\tilde{B}_n$

where

$$e^{\tilde{\Lambda}t} = \begin{pmatrix} e^{\lambda_1 t} & \cdots & 0 \\ & \ddots & \\ 0 & \cdots & e^{\lambda_n t} \end{pmatrix}$$

and $\tilde{B}_i$ are as defined in Equation (1.63).

Now let us consider matrix functions. The following are examples:

$$\tilde{A}(t) = \begin{pmatrix} t & 0 \\ t^2 & 4t \end{pmatrix} \quad \text{and} \quad \tilde{A}(\theta) = \begin{pmatrix} \sin\theta & \cos\theta \\ 0 & -\sin\theta \end{pmatrix}$$

We can easily extend the calculus of scalar functions to matrix functions. In the following, $\tilde{A}(t)$ and $\tilde{B}(t)$ are matrix functions with independent variable t and $\tilde{U}$ a matrix of real constants:

(i) $\dfrac{d}{dt}\tilde{U} = 0$

(ii) $\dfrac{d}{dt}(\alpha\tilde{A}(t) + \beta\tilde{B}(t)) = \alpha\dfrac{d}{dt}\tilde{A}(t) + \beta\dfrac{d}{dt}\tilde{B}(t)$

(iii) $\dfrac{d}{dt}(\tilde{A}(t)\tilde{B}(t)) = \left(\dfrac{d}{dt}\tilde{A}(t)\right)\tilde{B}(t) + \tilde{A}(t)\left(\dfrac{d}{dt}\tilde{B}(t)\right)$

(iv) $\dfrac{d}{dt}\tilde{A}^2(t) = \left(\dfrac{d}{dt}\tilde{A}(t)\right)\tilde{A}(t) + \tilde{A}(t)\left(\dfrac{d}{dt}\tilde{A}(t)\right)$

(v) $\dfrac{d}{dt}\tilde{A}^{-1}(t) = -\tilde{A}^{-1}(t)\left(\dfrac{d}{dt}\tilde{A}(t)\right)\tilde{A}^{-1}(t)$

(vi) $\dfrac{d}{dt}(\tilde{A}(t))^T = \left(\dfrac{d}{dt}\tilde{A}(t)\right)^T$

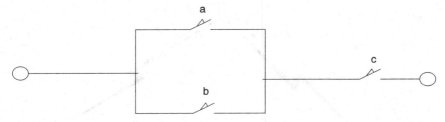

Figure 1.7 Switches for Problem 6

Problems

1. A pair of fair dice is rolled 10 times. What will be the probability that 'seven' will show at least once.
2. During Christmas, you are provided with two boxes A and B containing light bulbs from different vendors. Box A contains 1000 red bulbs of which 10% are defective while Box B contains 2000 blue bulbs of which 5% are defective.

   (a) If I choose two bulbs from a randomly selected box, what is the probability that both bulbs are defective?
   (b) If I choose two bulbs from a randomly selected box and find that both bulbs are defective, what is the probability that both came from Box A?

3. A coin is tossed an infinite number of times. Show that the probability that k heads are observed at the nth tossing but not earlier equals $\binom{n-1}{k-1} p^k (1-p)^{n-k}$, where $p = P\{H\}$.
4. A coin with $P\{H\} = p$ and $P\{T\} = q = 1 - p$ is tossed n times. Show that the probability of getting an even number of heads is $0.5[1 + (q - p)^n]$.
5. Let A, B and C be the events that switches a, b and c are closed, respectively. Each switch may fail to close with probability q. Assume that the switches are independent and find the probability that a closed path exists between the terminals in the circuit shown for $q = 0.5$.
6. The binary digits that are sent from a detector source generate bits 1 and 0 randomly with probabilities 0.6 and 0.4, respectively.

   (a) What is the probability that two 1 s and three 0 s will occur in a 5-digit sequence.
   (b) What is the probability that at least three 1 s will occur in a 5-digit sequence.

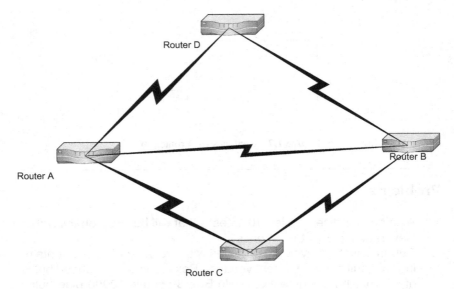

Router D

Router A

Router B

Router C

Figure 1.8 Communication network with 5 links

7. The binary input **X** to a channel takes on one of two values, 0 or 1, with probabilities 3/4 and 1/4 respectively. Due to noise induced errors, the channel output **Y** may differ from **X**. There will be no errors in **Y** with probabilities 3/4 and 7/8 when the input X is 1 or 0, respectively. Find $P(\mathbf{Y} = 1)$, $P(\mathbf{Y} = 0)$ and $P(\mathbf{X} = 1 \mid \mathbf{Y} = 1)$.

8. A wireless sensor node will fail sooner or later due to battery exhaustion. If the failure rate is constant, the time to failure T can be modelled as an exponentially distributed random variable. Suppose the wireless sensor node follow an exponential failure law in hours as $f_T(t) = \alpha\, u(t)e^{-\alpha t}$, where $u(t)$ is the unit step function and $\alpha > 0$ is a parameter. Measurements show that for these sensors, the probability that T exceeds 10^4 hours is e^{-1} (≈ 0.368). Using the value of the parameter α determined, calculate the time t_0 such that the probability that T is less than t_0 is 0.05.

9. A communication network consists of five links that interconnect four routers, as shown below. The probability that each of the link is operational is 0.9 and independent. What is the probability of being able to transmit a message from router A to router B (assume that packets move forward in the direction of the destination)?

10. Two random variables X and Y take on the values i and 2^i with probability $1/2^i$ ($i = 1, 2, \ldots$). Show that the probabilities sum to one. Find the expected value of X and Y.

11. There are three identical cards, one is red on both sides, one is yellow on both sides and the last one is red on one side and yellow on the other side. A card is selected at random and is red on the upper side. What is the probability that the other side is yellow?

2

Introduction to Queueing Systems

In today's information age society, where activities are highly interdependent and intertwined, sharing of resources and hence waiting in queues is a common phenomenon that occurs in every facet of our lives. In the context of data communication, expensive transmission resources in public data networks, such as the Internet, are shared by various network users. Data packets are queued in the buffers of switching nodes while waiting for transmission. In a computer system, computer jobs are queued for CPU or I/O devices in various stages of their processing.

The understanding and prediction of the stochastic behaviour of these queues will provide a theoretical insight into the dynamics of these shared resources and how they can be designed to provide better utilization. The modelling and analysis of waiting queues/networks and their applications in data communications is the main subject of this book. The study of queues comes under a discipline of *Operations Research* called *Queueing Theory* and is a primary methodological framework for evaluating resource performance besides simulation.

The origin of queueing theory can be traced back to early in the last century when A K Erlang, a Danish engineer, applied this theory extensively to study the behaviour of telephone networks. Acknowledged to be the father of queueing theory, Erlang developed several queueing results that still remain the backbone of queueing performance evaluations today.

In this chapter, you will be introduced to the basic structure, the terminology and the characteristics before embarking on the actual study of queueing

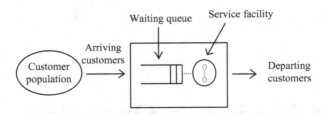

Figure 2.1   Schematic diagram of a queneing system

systems. The most commonly used Poisson process for modelling the arrival of customers to a queueing system will also be examined here. We assume that students are generally familiar with the basic college mathematics and probability theory, as refreshed in Chapter 1.

2.1 NOMENCLATURE OF A QUEUEING SYSTEM

In the parlance of queueing theory, a *queueing system* is a place where customers arrive according to an 'arrival process' to obtain service from a service facility. The service facility may contain more than one server (or more generally resources) and it is assumed that a server can serve one customer at a time. If an arriving customer finds all servers occupied, he joins a waiting queue. This customer will receive his service later, either when he reaches the head of the waiting queue or according to some service discipline. He leaves the system upon completion of his service.

The schematic diagram of a queueing system is depicted in Figure 2.1. A queueing system is sometimes just referred to as a queue, or queueing node (or node for short) in some queueing literature. We may use these terms interchangeably when we discuss networks of queueing systems in the sequel when there is unlikely to be any confusion.

In the preceding description, we used the generic terms 'customers' and 'servers', which are in line with the terms used in queueing literature. They take various forms in the different domains of applications. In the case of a data switching network, 'customers' are data packets (or data frames or data cells) that arrive at a switching node and 'servers' are the transmission channels. In a CPU job scheduling problem, 'customers' are computer processes (jobs or transactions) and 'servers' are the various computer resources, such as CPU, I/O devices.

So given such a dynamic picture of a queueing system, how do we describe it analytically? How do we formulate a mathematical model that reflects these dynamics? What are the parameters that characterize a queueing system completely? Before we proceed let us examine the structure of a queueing system. Basically, a queueing system consists of three major components:

- The input process
- The system structure
- The output process.

2.1.1   Characteristics of the Input Process

When we talk about the input process, we are in fact concerning ourselves with the following three aspects of the arrival process:

(1) The size of arriving population
The size of the arriving customer population may be infinite in the sense that the number of potential customers from external sources is very large compared to those in the system, so that the arrival rate is not affected by the size.

The size of the arriving population has an impact on the queueing results. An infinite population tends to render the queueing analysis more tractable and often able to provide simple closed-form solutions, hence this model will be assumed for our subsequent queueing systems unless otherwise stated. On the other hand, the analysis of a queueing system with finite customer population size is more involved because the arrival process is affected by the number of customers already in the system.

Examples of infinite customers populations are telephone users requesting telephone lines, and air travellers calling an air ticket reservation system. In fact, these are actually finite populations but they are very large, and for mathematical convenience we treat them as infinite. Examples of the finite calling populations would be a group of stations in a local area network presenting data frame to the broadcast channel, or a group of video display units requesting response from a CPU.

(2) Arriving patterns
Customers may arrive at a queueing system either in some regular pattern or in a totally random fashion. When customers arrive regularly at a fixed interval, the arriving pattern can be easily described by a single number – the rate of arrival. However, if customers arrive according to some random mode, then we need to fit a statistical distribution to the arriving pattern in order to render the queueing analysis mathematically feasible.

The parameter that we commonly use to describe the arrival process is the inter-arrival time between two customers. We generally fit a probability distribution to it so that we can call upon the vast knowledge of probability theory. The most commonly assumed arriving pattern is the Poisson process whose inter-arrival times are exponentially distributed. The popularity of the Poisson process lies in the fact that it describes very well a completely random arrival pattern, and also leads to very simple and elegant queueing results.

We list below some probability distributions that are commonly used to describe the inter-arrival time of an arrival process. These distributions are generally denoted by a single letter as shown:

M: Markovian (or Memoryless), imply Poisson process
D: Deterministic, constant interarrival times
E_k: Erlang distribution of order K of inter-arrival times
G: General probability distribution of inter-arrival times
GI: General and independent (inter-arrival time) distribution.

(3) Behaviour of arriving customers
Customers arriving at a queueing system may behave differently when the system is full due to a finite waiting queue or when all servers are busy. If an arriving customer finds the system is full and leaves forever without entering the system, that queueing system is referred to as a blocking system. The analysis of blocking systems, especially in the case of queueing networks, is more involved and at times it is impossible to obtain closed-form results. We will assume that this is the behaviour exhibited by all arriving customers in our subsequent queueing models. In real life, customers tend to come back after a short while.

In telephone networks, call requests that are blocked when the network is busy are said to employ a lost-calls-cleared discipline. The key measure of performance in such a system is the probability of blocking that a call experiences. We will discuss blocking probability in greater detail in Chapter 4. On the other hand, if calls are placed in queues, it is referred to as lost-calls-delayed. The measure of performance is the elapsed time of a call.

2.1.2   Characteristics of the System Structure

(i) Physical number and layout of servers
The service facility shown in Figure 2.1 may comprise of one or more servers. In the context of this book, we are interested only in the parallel and identical servers; that is a customer at the head of the waiting queue can go to any server that is free, and leave the system after receiving his/her service from that server, as shown in Figure 2.2 (a). We will not concern ourselves with the serial-servers case where a customer receives services from all or some of them in stages before leaving the system, as shown in Figure 2.2 (b).

(ii) The system capacity
The system capacity refers to the maximum number of customers that a queueing system can accommodate, inclusive of those customers at the service facility.

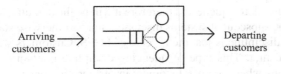

Figure 2.2 (a)   Parallel servers

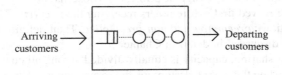

Figure 2.2 (b)   Serial servers

In a multi-server queueing system, as shown in Figure 2.2 (a), the system capacity is the sum of the maximum size of the waiting queue and the number of servers. If the waiting queue can accommodate an infinite number of customers, then there is no blocking, arriving customers simply joining the waiting queue. If the waiting queue is finite, then customers may be turned away. It is much easier to analyse queueing systems with infinite system capacity as they often lead to power series that can be easily put into closed form expressions.

2.1.3   Characteristics of the Output Process

Here, we are referring to the following aspects of the service behaviour as they greatly influence the departure process.

(i) Queueing discipline or serving discipline
Queueing discipline, sometimes known as serving discipline, refers to the way in which customers in the waiting queue are selected for service. In general, we have:

- First-come-first-served (FCFS)
- Last-come-first-served (LCFS)
- Priority
- Processor sharing
- Random.

The FCFS queueing discipline does not assign priorities and serves customers in the order of their arrivals. Apparently this is the most frequently encountered discipline at an ordered queue and therefore it will be the default queueing discipline for all the subsequent systems discussed, unless otherwise stated.

The LCFS discipline is just the reverse of FCFS. Customers who come last will be served first. This type of queueing discipline is commonly found in stack operations where items (customers in our terminology) are stacked and operations occur only at the top of the stack.

In priority queueing discipline, customers are divided into several priority classes according to their assigned priorities. Those having a higher priority than others are served first before others receiving their service. There are two sub-classifications: *preemptive* and *non-preemptive*. Their definitions and operations will be discussed in detail in Chapter 5.

In processor sharing, capacity is equally divided among all customers in the queue, that is when there are k customers, the server devotes $1/k$ of his capacity to each. Equally, each customer obtains service at $1/k$ of rate and leaves the system upon completion of his service.

(ii) Service-time distribution
Similar to arrival patterns, if all customers require the same amount of service time then the service pattern can be easily described by a single number. But generally, different customers require different amounts of service times, hence we again use a probability distribution to describe the length of service times the server renders to those customers. The most commonly assumed service time distribution is the negative exponential distribution.

Again, we commonly use a single letter to indicate the type of service distributions:

M: Markovian (or Memoryless) , imply exponential distributed service times
D: Deterministic ; constant service times
E_k: Erlang of order K service time distribution
G: General service times distribution.

2.2   RANDOM VARIABLES AND
THEIR RELATIONSHIPS

From the preceding description of a queueing system, we see that customers arrive, are served and leave the system, hence presenting a fluid situation with constant motions. There are many processes present and interacting with each other. Most of the quantities associated with these processes evolve in time and are of a probabilistic nature. In other words, they are the so-called random variables and their values can only be expressed through probability. We summarize the primary random variables of a queueing system in Table 2.1 and list some of the relationships among them.

Table 2.1 Random variables of a queueing system

Notation	Description
$N(t)$	The number of customers in the system at time t
$N_q(t)$	The number of customers in the waiting queue at time t
$N_s(t)$	The number of customers in the service facility at time t
N	The average number of customers in the system
N_q	The average number of customers in the waiting queue
N_s	The average number of customers in the service facility
T_k	The time spent in the system by kth customer
W_k	The time spent in the waiting queue by kth customer
x_k	The service time of kth customer
T	The average time spent in the system by a customer
W	The average time spent in the waiting queue by a customer
$\bar{x}$	The average service time
$P_k(t)$	The probability of having k customers in the system at time t
P_k	The stationary probability of having k customers in the system

Looking at the structure of a queueing system, we can easily arrive at the following expressions:

$$N(t) = N_q(t) + N_s(t) \quad \text{and} \quad N = N_q + N_s \qquad (2.1)$$

$$T_k = W_k + x_k \quad \text{and} \quad T = W + \bar{x} \qquad (2.2)$$

If the queueing system in question is ergodic (a concept that we shall explain later in Section 2.4.2) and has reached the steady state, then the following expressions hold:

$$N = \lim_{t \to \infty} N(t) = \lim_{t \to \infty} E[N(t)] \qquad (2.3)$$

$$N_q = \lim_{t \to \infty} N_q(t) = \lim_{t \to} E[N_q(t)] \qquad (2.4)$$

$$N_s = \lim_{t \to \infty} N_s(t) = \lim_{t \to \infty} E[N_s(t)] \qquad (2.5)$$

$$T = \lim_{k \to \infty} T_k = \lim_{k \to \infty} E[T_k] \qquad (2.6)$$

$$W = \lim_{k \to \infty} W_k = \lim_{k \to \infty} E[W_k] \qquad (2.7)$$

$$\bar{x} = \lim_{k \to \infty} x_k = \lim_{k \to \infty} E[x_k] \qquad (2.8)$$

$$P_k = \lim_{t \to \infty} P_k(t) \qquad (2.9)$$

2.3   KENDALL NOTATION

From the above section, we see that there are many stochastic processes and a multiplicity of parameters (random variables) involved in a queueing system, so given such a complex situation how do we categorize them and describe them succinctly in a mathematical short form? David G Kendall, a British statistician, devised a shorthand notation, shown below, to describe a queueing system containing a single waiting queue. This notation is known as Kendall notation:

$$A / B / X / Y / Z$$

where   A : Customer arriving pattern (Inter-arrival-time distribution)
        B : Service pattern (Service-time distribution)
        X : Number of parallel servers
        Y : System capacity
        Z : Queueing discipline

For example, M / M / 1 / ∞ / FCFS represents a queuing system where customers arrive according to a Poisson process and request exponentially distributed service times from the server. The system has only one server, an infinite waiting queue and customers are served on an FCFS basis. In many situations, we only use the first three parameters, for example, M / D / 1. The default values for the last two parameters are Y = ∞ and Z = FCFS.

Example 2.1: Modelling of a job-processing computer system

Figure 2.3 shows a computer setup where a pool of m remote-job-entry terminals is connected to a central computer. Each operator at the terminals spends an average of S seconds thinking and submitting a job (or request) that requires P seconds at the CPU. These submitted jobs are queued and later processed by the single CPU in an unspecified queueing discipline.

We would like to estimate the maximum throughput of the system, so propose a queueing model that allows us to do so.

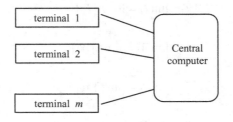

Figure 2.3   A job-processing system

Solution

There are two ways of modelling this terminal-CPU system.

(i) Firstly, we assume that the thinking times (or job submission times) and CPU processing times are exponentially distributed, hence the arrival process is Poisson. Secondly, we further assume that those operators at the terminals are either waiting for their responses from the CPU or actively entering their requests, so if there are k requests waiting to be processed there are $(m - k)$ terminals active in the arrival process. Thus we can represent the jobs entered as a state-dependent Poisson process with the rate:

$$\lambda(k) = \begin{cases} (m-k)\lambda & k < m \\ 0 & k \geq m \end{cases} \tag{2.10}$$

Therefore, the system can be modelled as a M/M/1/m system, as shown in Figure 2.4, with arrival rate $\lambda(k)$ given by expression (2.10) above. This model is commonly known as a 'finite population' queueing model or 'machine repairman' model in classical queueing theory.

(ii) An alternate way is to model the system as a closed queueing network of two queues, as shown in Figure 2.5; one for the terminals and the other for the CPU. Closed queueing networks will be discussed in detail in Chapter 7.

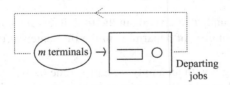

Figure 2.4   An M/M1/m with finite customer population

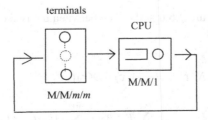

Figure 2.5   A closed queueing netwook model

2.4   LITTLE'S THEOREM

Before we examine the stochastic behaviour of a queueing system, let us first establish a very simple and yet powerful result that governs its steady-state performance measures – *Little's theorem, law* or *result* are the various names. This result existed as an empirical rule for many years and was first proved in a formal way by J D C Little in 1961 (Little 1961).

This theorem relates the average number of customers (N) in a steady-state queueing system to the product of the average arrival rate (λ) of customers entering the system and the average time (T) a customer spent in that system, as follows:

$$N = \lambda T \qquad\qquad (2.11)$$

This result was derived under very general conditions. The beauty of it lies in the fact that it does not assume any specific distribution for the arrival as well as the service process, nor it assumes any queueing discipline or depends upon the number of parallel servers in the system. With proper interpretation of N, λ and T, it can be applied to all types of queueing systems, including priority queueing and multi-server systems.

Here, we offer a simple proof of the theorem for the case when customers are served in the order of their arrivals. In fact, the theorem holds for any queueing disciplines as long as the servers are kept busy when the system is not empty.

Let us count the number of customers entering and leaving the system as functions of time in the interval $(0, t)$ and define the two functions:

$A(t)$: Number of arrivals in the time interval $(0, t)$
$D(t)$: Number of departures in the time interval $(0, t)$

Assuming we begin with an empty system at time 0, then $N(t) = A(t) - D(t)$ represents the total number of customers in the system at time t. A general sample pattern of these two functions is depicted in Figure 2.6, where t_k is the instant of arrival and T_k the corresponding time spent in the system by the kth customer.

It is clear from Figure 2.6 that the area between the two curves $A(t)$ and $D(t)$ is given by

$$\int_0^t N(\tau)d\tau = \int_0^t [A(\tau) - D(\tau)]d\tau$$
$$= \sum_{k=1}^{D(t)} T_k \times 1 + \sum_{k=D(t)+1}^{A(t)} (t - t_k) \times 1$$

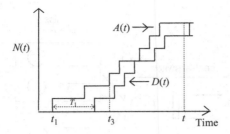

Figure 2.6 A sample pattern of arrival

hence
$$\frac{1}{t}\int_0^t N(\tau)d\tau = \left[\sum_{k=1}^{D(t)} T_k + \sum_{k=D(t)+1}^{A(t)} (t-t_k) \right] \times \frac{1}{A(t)} \times \frac{A(t)}{t} \qquad (2.12)$$

We recognize that the left-hand side of the equation is simply the *time-average* of the number of customers (N_t) in the system during the interval $(0, t)$, whereas the last term on the right $A(t)/t$ is the time-average of the customer arrival rate (λ_t) in the interval $(0, t)$. The remaining bulk of the expression on the right:

$$\left[\sum_{i=1}^{D(t)} T_i + \sum_{i=D(t)+1}^{A(t)} (t-t_i) \right] \Big/ A(t)$$

is the time-average of the time (T_t) a customer spends in the system in $(0, t)$. Therefore, we have

$$N_t = \lambda_t \times T_t \qquad (2.13)$$

Students should recall from probability theory that the number of customers in the system can be computed as

$$N = \sum_{k=0}^{\infty} k P_k$$

The quantity N computed in this manner is the *ensemble* (or *stochastic*) *average* or the so-called expected value. In Section 2.2 we mentioned that ensemble averages are equal to time averages for an ergodic system, and in general most of the practical queueing systems encountered are ergodic, thus we have

$$N = \lambda \times T \qquad (2.14)$$

where λ and T are ensemble quantities.

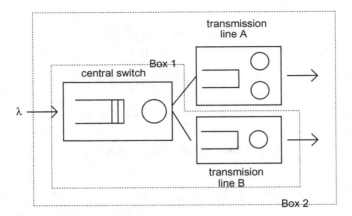

Figure 2.7   A queueing model of a switch

This result holds for both the time average quantities as well as stochastic (or ensemble) average quantities. We shall clarify further the concept of ergodicity later and we shall assume that all queueing systems we are dealing with are ergodic in subsequent chapters.

2.4.1   General Applications of Little's Theorem

Little's formula was derived in the preceding section for a queuing system consisting of only one server. It is equally valid at different levels of abstraction, for example the waiting queue, or a larger system such as a queueing network. To illustrate its applications, consider the hypothetical queueing network model of a packet switching node with two outgoing links, as depicted in Figure 2.7.

If we apply Little's result to the waiting queue and the service facility of the central switch respectively, we have

$$N_q = \lambda W \tag{2.15}$$

$$N_s = \lambda \bar{x} = \rho \tag{2.16}$$

where N_q is the number of packets in the waiting queue and N_s is the average number of packets at the service facility. We introduce here a new quantity ρ, a utilization factor, which is always less than one for a stable queueing system. We will say more about this in Section 2.5.

Box 1 illustrates the application to a sub-system which comprises the central switch and transmission line B. Here the time spent in the system (T_{sub}) is the total time required to traverse the switch and the transmission line B. The number of packets in the system (N_{sub}) is the sum of packets at both the switch and the transmission line B:

$$N_{sub} = N_{sw} + N_B = \lambda T_{sw} + \lambda_B T_B$$

$$T_{sub} = T_{sw} + T_B \tag{2.17}$$

where T_{sw} is the time spent at the central switch
         T_B is the time spent by a packet to be transmitted over Line B
         N_{sw} is the number of packets at the switch
         N_B is the number of packets at Line B
         λ_B is the arrival rate of packets to Line B.

We can also apply Little's result to the system as a whole (Box 2). Then T and N are the corresponding quantities of that system:

$$N = \lambda T \tag{2.18}$$

2.4.2   Ergodicity

Basically, '*Ergodicity*' is a concept related to the limiting values and stationary distribution of a long sample path of a stochastic process. There are two ways of calculating the average value of a stochastic process over a certain time interval. In experiments, the average value of a stochastic process $X(t)$ is often obtained by observing the process for a sufficiently long period of time $(0, T)$ and then taking the average as

$$\overline{X} = \frac{1}{T} \int_0^T X(t)dt$$

The average so obtained is called the time average. However, a stochastic process may take different realizations during that time interval, and the particular path that we observed is only a single realization of that process. In probability theory, the expected value of a process at time T refers to the average of the various realizations at time T, as shown in Figure 2.8, and is given by

$$E[X(T)] = \sum_{k=0}^{\infty} kP[X(T) = k]$$

This expected value is also known as the ensemble average. For a stochastic process, if the time average is equal to the ensemble average as $T \to \infty$, we say that the process is ergodic.

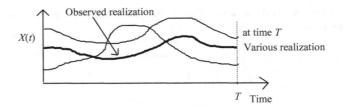

Figure 2.8   Ensemble average of a process

2.5   RESOURCE UTILIZATION AND TRAFFIC INTENSITY

The concept of resource utilization, sometimes referred to as utilization factor (or just utilization), is basically a measure of how busy a resource is. In the context of queueing theory, it represents the fraction of time a server is engaged in providing service, defined as

$$\text{Utilization } (\rho) = \frac{\text{Time a server is occupied}}{\text{Time available}}$$

In the case of a queueing system with $m(m \geq 1)$ servers, if there are N customers in the system within a time interval $(t, t + T)$, then each server on average will serve $(\lambda T)/m$ customers. If the average service time is $\bar{x} = 1/\mu$ unit of time, then we have

$$\rho = \frac{(\lambda T/m) \times (1/\mu)}{T} = \frac{\lambda}{m\mu} \tag{2.19}$$

It is clear from the above definition that ρ is dimensionless and should be less than unity in order for a server to cope with the service demand, or in other words for the system to be stable.

A measure that is commonly used in traffic engineering and closely related to the concept of resource utilization is *traffic intensity*. It is a measure of the total arrival traffic presented to the system as a whole. It is also known as *offered load* and is by definition just the product of average arrival rate and the average service time:

$$\text{Traffic intensity } (\alpha) = \lambda\bar{x} = \lambda/\mu \tag{2.20}$$

In fact, traffic intensity is again a dimensionless quantity but it is usually expressed in erlangs, named after A K Erlang.

To better understand the physical meaning of this unit, take a look at the traffic presented to a single resource. One erlang of traffic is equivalent to a single user who uses that resource 100% of the time or alternatively, 10 users who each occupy the resource 10% of the time. A traffic intensity greater than one indicates that customers arrive faster than they are served and is a good indication of the minimum number of servers required to achieve a stable system. For example, a traffic intensity of 2.5 erlangs indicates that at least 3 servers are required.

2.6 FLOW CONSERVATION LAW

This is analogous to Kirchhoff's current law which states that the algebraic sum of all the electrical currents entering a node must equal the algebraic sum of all the currents leaving a node.

For a stable queueing system, the rate of customers entering the system should equal the rate of customers leaving the system, if we observe it for a sufficiently long period of time, i.e. $\lambda_{out} = \lambda_{in}$, as shown in Figure 2.9. This is initiatively clear because if $\lambda_{out} < \lambda_{in}$ then there will be a steady build-up of customers and the system will eventually become unstable. On the other hand, if $\lambda_{out} > \lambda_{in}$, then customers are created within the system.

This notion of flow conservation is useful when we wish to calculate throughput. It can be applied to individual queued in a collection of queueing systems. We will see in Chapter 6 how it is used to derive traffic equations for each of the queues in a network of queues.

Example 2.2

(a) A communication channel operating at 9600 bps receives two types of packet streams from a gateway. Type A packets have a fixed length format of 48 bits whereas Type B packets have an exponentially distributed length with a mean of 480 bits. If on average there are 20% Type A packets and 80% Type B packets, calculate the utilization of this channel assuming the combined arrival rate is 15 packets per second.
(b) A PBX was installed to handle the voice traffic generated by 300 employees in a factory. If each employee, on average, makes 1.5 calls per hour with

Figure 2.9 Flow Conservation Law

an average call duration of 2.5 minutes, what is the offered load presented
to this PBX?

Solution

(a) The average transmission time
$$= (0.2 \times 48 + 0.8 \times 480)/9600$$
$$= 0.041 \text{ s}$$
$$\rho = 15 \times 0.041 = 61.5\%$$

(b) *Offered load = Arrival rate × Service time*
$$= 300 \ (users) \times 1.5 \ (calls \ per \ user \ per \ hour)$$
$$\times 2.5 \ (minutes \ percall) \div 60 \ (minutes \ perhour)$$
$$= 18.75 \ erlangs$$

Example 2.3

Consider the queueing system shown below where we have customers arriving
to Queue 1 and Queue 2 with rates γ_a and γ_b, respectively. If the branching
probability p at Queue 2 is 0.5, calculate the effective arrival rates to both
queues.

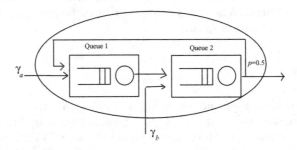

Solution

Denote the effective arrival rates to Queue 1 and Queue 2 as λ_1 and λ_2, respectively, then we have the following expressions under the principal of flow
conservation:

$$\lambda_1 = \gamma_a + 0.5\lambda_2$$

$$\lambda_2 = \lambda_1 + \gamma_b$$

Hence, we have

$$\lambda_1 = 2\gamma_a + \gamma_b$$
$$\lambda_2 = 2(\gamma_a + \gamma_b)$$

2.7 POISSON PROCESS

Poisson process is central to physical process modelling and plays a pivotal role in classical queuing theory. In most elementary queueing systems, the inter-arrival times and service times are assumed to be exponentially distributed or, equivalently, that the arrival and service processes are Poisson, as we shall see below. The reason for its ubiquitous use lies in the fact that it possesses a number of marvellous probabilistic properties that give rise to many elegant queueing results. Secondly, it also closely resembles the behaviour of numerous physical phenomenon and is considered to be a good model for an arriving process that involves a large number of similar and independent users.

Owing to the important role of Poisson process in our subsequent modelling of arrival processes to a queueing system, we will take a closer look at it and examine here some of its marvellous properties.

Put simply, a Poisson process is a counting process for the number of randomly occurring point events observed in a given time interval $(0, t)$. It can also be deemed as the limiting case of placing at random k points in the time interval of $(0, t)$. If the random variable $X(t)$ that counts the number of point events in that time interval is distributed according to the well-known Poisson distribution given below, then that process is a Poisson process:

$$P[X(t) = k] = \frac{(\lambda t)^k}{k!} e^{-\lambda t} \tag{2.21}$$

Here, λ is the rate of occurrence of these point events and λt is the mean of a Poisson random variable and physically it represents the average number of occurrences of the event in a time-interval t. Poisson distribution is named after the French mathematician, Simeon Denis Poisson.

2.7.1 The Poisson Process – A Limiting Case

The Poisson process can be considered as a limiting case of the Binomial distribution of a Bernoulli trial. Assuming that the time interval $(0, t)$ is divided into time slots and each time slot contains only one point, if we place points at random in that interval and consider a point in a time slot as a 'success', then the number of k 'successes' in n time slots is given by the Binomial distribution:

$$P[k \text{ successes in } n \text{ time-slots}] = \binom{n}{k} p^k (1-p)^{n-k}$$

Now let us increase the number of time slots (n) and at the same time decrease the probability (p) of 'success' in such a way that the average number of 'successes' in a time interval t remains constant at $np = \lambda t$, then we have the Poisson distribution:

$P[k \text{ arrivals in } (0, t)]$

$$= \lim_{n \to \infty} \binom{n}{k} \left(\frac{\lambda t}{n} \right)^k \left(1 - \frac{\lambda t}{n} \right)^{n-k}$$

$$= \frac{(\lambda t)^k}{k!} \left[\lim_{n \to \infty} \frac{n(n-1) \dots (n-k+1)}{n^k} \right] \left[\lim_{n \to \infty} \left(1 - \frac{\lambda t}{n} \right)^n \right]$$

$$= \frac{(\lambda t)^k}{k!} \left[\lim_{n \to \infty} \left\{ \left(1 - \frac{\lambda t}{n} \right)^{-\frac{n}{\lambda t}} \right\}^{-\lambda t} \right]$$

$$= \frac{(\lambda t)^k}{k!} e^{-\lambda t} \qquad\qquad (2.22)$$

In the above expression, we made use of the identity:

$$e = \lim_{n \to \infty} \left(1 - \frac{\lambda t}{n} \right)^{-n/\lambda t}$$

So we see that the process converges to a Poisson process. Therefore, Poisson process can be deemed as the superposition of a large number of Bernoulli arrival processes.

2.7.2   The Poisson Process – An Arrival Perspective

Poisson process has several interpretations and acquires different perspectives depending on which angle we look at it. It can also be viewed as a counting process $\{A(t), t \geq 0\}$ which counts the number of arrivals up to time t where $A(0) = 0$. If this counting process satisfies the following assumptions then it is a Poisson process.

(i) The distribution of the number of arrivals in a time interval depends only on the length of that interval. That is

$$P[A(\Delta t) = 0] = 1 - \lambda \Delta t + o(\Delta t)$$
$$P[A(\Delta t) = 1] = \lambda \Delta t + o(\Delta t)$$
$$P[A(\Delta t) \geq 2] = o(\Delta t)$$

where $o(\Delta t)$ is a function such that $\lim\limits_{\Delta t \to 0} \dfrac{o(\Delta t)}{\Delta t} = 0$.

This property is known as independent increments.

(ii) The number of arrivals in non-overlapping intervals is statistically independent. This property is known as stationary increments.

The result can be shown by examining the change in probability over a time interval $(t, t + \Delta t)$. If we define $P_k(t)$ to be the probability of having k arrivals in a time interval t and examine the change in probability in that interval, then

$$\begin{aligned} P_k(t + \Delta t) &= P[k\ arrivals\ in\ (0, t + \Delta t)] \\ &= \sum_{i=0}^{k} P[(k-i)\ in\ (0, t)\ \&\ i\ in\ \Delta t] \end{aligned} \tag{2.23}$$

Using the assumptions (i) and (ii) given earlier:

$$\begin{aligned} P_k(t + \Delta t) &= \sum_{i=0}^{k} P[(k-i)\ arrivals\ in\ (0, t)] \times P[i\ in\ \Delta t] \\ &= P_k(t)[1 - \lambda \Delta t + 0(\Delta t)] + P_{k-1}(t)[\lambda \Delta t + 0(\Delta t)] \end{aligned} \tag{2.24}$$

Rearranging the terms, we have

$$P_k(t + \Delta t) - P_k(t) = -\lambda \Delta t P_k(t) + \lambda \Delta t P_{k-1}(t) + O(\Delta t)$$

Dividing both equations by Δt and letting $\Delta t \to 0$, we arrive at

$$\frac{dP_k(t)}{dt} = -\lambda P_k(t) + \lambda P_{k-1}(t) \quad k > 0 \tag{2.25}$$

Using the same arguments, we can derive the initial condition:

$$\frac{dP_0(t)}{dt} = -\lambda P_0(t) \tag{2.26}$$

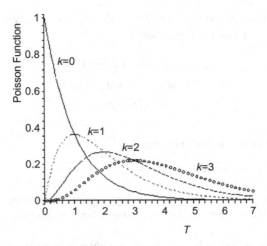

Figure 2.10   Sample Poisson distributuon Function

From the example in Chapter 1, we see that the solution to these two differential equations is given by

$$P_0(0) = e^{-\lambda t} \tag{2.27}$$

$$P_k(t) = \frac{(\lambda t)^k}{k!} e^{-\lambda t} \tag{2.28}$$

which is the Poisson distribution. A sample family of Poisson distributions for $k = 0, 1, 2$ *and* 3 are plotted in Figure 2.10. The x-axis $T = \lambda t$.

2.8   PROPERTIES OF THE POISSON PROCESS

2.8.1   Superposition Property

The superposition property says that if k independent Poisson processes A_1, $A_2, \ldots A_k$ are combined into a single process $A = A_1 + A_2 + \ldots + A_k$, then A is still Poisson with rate λ equal to the sum of the individual rates λ_i of A_i, as shown in Figure 2.11.

Recall that the z-transform of a Poisson distribution with parameter λt is $e^{-\lambda t(1-z)}$.

Since $A = A_1 + A_2 + \cdots + A_k$

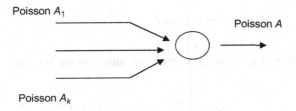

Figure 2.11 Superposition property

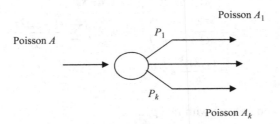

Figure 2.12 Decomposition property

Taking expectations of both sides, we have

$$
\begin{aligned}
E\left[z^{A}\right] &= E\left[Z^{A_1+A_2+\cdots+A_K}\right] \\
&= E\left[z^{A_1}\right]E\left[z^{A_2}\right]\ldots E\left[z^{A_k}\right] \quad \text{(indenpendence assumption)} \\
&= e^{-\lambda_1 t(1-z)}\ldots e^{-\lambda_k t(1-z)} \\
&= e^{-(\lambda_1+\lambda_2+\cdots+\lambda_k)t(1-z)}
\end{aligned}
\tag{2.29}
$$

The right-hand side of the final expression is the z-transform of a Poisson distribution with rate $(\lambda_1 + \lambda_2 + \ldots + \lambda_k)$, hence the resultant process is Poisson.

2.8.2 Decomposition Property

The decomposition property is just the reverse of the previous property, as shown in Figure 2.12, where a Poisson process A is split into k processes using probability p_i ($i = 1, \ldots, k$).

Let us derive the probability mass function of a typical process A_i. On condition that there are N arrivals during the time interval $(0, t)$ from process A, the probability of having k arrivals at process A_i is given by

$$P[A_i(t) = k | A(t) = N \ \& \ N \geq k] = \binom{N}{k} P_i^k (1 - p_i)^{N-k} \qquad (2.30)$$

The unconditional probability is then calculated using the total probability theorem:

$$
\begin{aligned}
P[A_i(t) = k] &= \sum_{N=k}^{\infty} \frac{N!}{(N-k)!k!} p_i^k (1 - p_i)^{N-k} \frac{(\lambda t)^N}{N!} e^{-\lambda t} \\
&= \frac{e^{-\lambda t}}{k!} \left(\frac{p_i}{1 - p_i} \right)^k \sum_{N=k}^{\infty} \frac{[(1 - p_i)\lambda t]^N}{(N-k)!} \\
&= \frac{e^{-\lambda t}}{k!} \left(\frac{p_i}{1 - p_i} \right)^k [(1 - p_i)\lambda t]^k \sum_{j=0}^{\infty} \frac{[(1 - p_i)\lambda t]^j}{j!} \\
&= \frac{(p_i \lambda t)^k}{k!} e^{-p_i \lambda t} \qquad (2.31)
\end{aligned}
$$

That is, a Poisson process with rate $p_i \lambda$.

2.8.3 Exponentially Distributed Inter-arrival Times

The exponential distribution and the Poisson process are closely related and in fact they mirror each other in the following sense. If the inter-arrival times in a point process are exponentially distributed, then the number of arrival points in a time interval is given by the Poisson distribution and the process is a Poisson arrival process. Conversely, if the number of arrival points in any interval is a Poisson random variable, the inter-arrival times are exponential distributed and the arrival process is Poisson.

Let τ be the inter-arrival time, then

$$P[\tau \leq t] = 1 - P[\tau > t].$$

But $P[\tau > t]$ is just the probability that no arrival occurs in $(0, t)$; i.e. $P_0(t)$. Therefore we obtain

$$P[\tau \leq t] = 1 - e^{-\lambda t} \quad \text{(exponential distribution)} \qquad (2.32)$$

2.8.4 Memoryless (Markovian) Property of Inter-arrival Times

The memoryless property of a Poisson process means that if we observe the process at a certain point in time, the distribution of the time until next arrival is not affected by the fact that some time interval has passed since the last

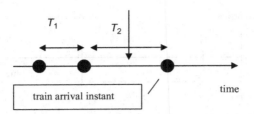

Figure 2.13 Sample train arrival instants

arrival. In other words, the process starts afresh at the time of observation and has no memory of the past. Before we deal with the formal definition, let us look at an example to illustrate this concept.

Example 2.4

Consider the situation where trains arrive at a station according to a Poisson process with a mean inter-arrival time of 10 minutes. If a passenger arrives at the station and is told by someone that the last train arrived 9 minutes ago, so on the average, how long does this passenger need to wait for the next train?

Solution

Intuitively, we may think that 1 minute is the answer, but the correct answer is 10 minutes. The reason being that Poisson process, and hence the exponential inter-arrival time distribution, is memoryless. What have happened before were sure events but they do not have any influence on future events.

This apparent 'paradox' lies in the renewal theory and can be explained qualitatively as such. Though the average inter-arrival time is 10 minutes, if we look at two intervals of inter-train arrival instants, as shown in Figure 2.13, a passenger is more likely to arrive within a longer interval T_2 rather than a short interval of T_1. The average length of the interval in which a customer is likely to arrive is twice the length of the average inter-arrival time. We will re-visit this problem quantitatively in Chapter 5.

Mathematically, the 'memoryless' property states that the distribution of remaining time until the next arrival, given that t_0 units of time have elapsed since the last arrival, is identically equal to the unconditional distribution of inter-arrival times (Figure 2.14).

Assume that we start observing the process immediately after an arrival at time 0. From Equation (2.21) we know that the probability of no arrivals in (0, t_0) is given by

$$P[no\ arrival\ in\ (0,\ t_0)] = e^{-\lambda t_0}$$

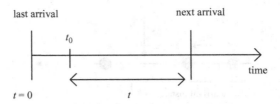

Figure 2.14   Conditional inter-arrival times

Let us now find the conditional probability that the first arrival occurs in $[t_0, t_0 + t]$, given that t_0 has elapsed; that is

$$P[arrival\ in\ (t_0, t_0 + t)|no\ arrival\ in(0, t_0)] = \frac{\int_{t_0}^{t_0+t} \lambda e^{-\lambda t}}{e^{-\lambda t_0}} = 1 - e^{-\lambda t}$$

But the probability of an arrival in $(0, t)$ is also $\int_0^t \lambda e^{-\lambda t} dt = 1 - e^{-\lambda t}$

Therefore, we see that the conditional distribution of inter-arrival times, given that certain time has elapsed, is the same as the unconditional distribution. It is this memoryless property that makes the exponential distribution ubiquitous in stochastic models. Exponential distribution is the only continuous function that has this property; its discrete counterpart is the geometry distribution.

2.8.5   Poisson Arrivals During a Random Time Interval

Consider the number of arrivals (N) in a random time interval I. Assuming that I is distributed with a probability density function $A(t)$ and I is independent of the Poisson process, then

$$P(N = k) = \int_0^\infty P(N = k|I = t) A(t) dt$$

But

$$P(N = k|I = t) = \frac{(\lambda t)^k}{k!} e^{-\lambda t}$$

Hence

$$P(N = k) = \int_0^\infty \frac{(\lambda t)^k}{k!} e^{-\lambda t} A(t) dt$$

Taking the z-transform, we obtain

$$N(z) = \sum_{k=0}^{\infty}\left(\int_0^{\infty}\frac{(\lambda t)^k}{k!}e^{-\lambda t}A(t)dt \right)z^k$$

$$= \int_0^{\infty}e^{-\lambda t}\sum_{k=0}^{\infty}\frac{(\lambda tz)^k}{k!}A(t)dt$$

$$= \int_0^{\infty}e^{-(\lambda-\lambda z)t}A(t)dt$$

$$= A^*(\lambda - \lambda z)$$

where $A^*(\lambda - \lambda z)$ is the Laplace transform of the arrival distribution evaluated at the point $(\lambda - \lambda z)$.

Example 2.5

Let us consider again the problem presented in Example 2.4. When this passenger arrives at the station:

a) What is the probability that he will board a train in the next 5 minutes?
b) What is the probability that he will board a train in 5 to 9 minutes?

Solution

a) From Example 2.4, we have $\lambda = 1/10 = 0.1$ min^{-1}, hence for a time period of 5 minutes we have

$$\lambda t = 5 \times 0.1 = 0.5 \quad \text{and}$$

$$P[0 \text{ } train \text{ } in \text{ } 5\min] = \frac{e^{-\lambda t}(\lambda t)^k}{k!} = \frac{e^{-0.5}(0.5)^0}{0!} = 0.607$$

He will board a train if at least one train arrives in 5 minutes; hence

$$P[at \text{ } least \text{ } 1 \text{ } train \text{ } in \text{ } 5 \text{ } min] = 1 - P[0 \text{ } train \text{ } in \text{ } 5 \text{ } min]$$
$$= 0.393$$

b) He will need to wait from 5 to 9 minutes if no train arrives in the first 5 minutes and board a train if at least one train arrives in the time interval 5 to 9 minutes. From (a) we have

$$P[0 \text{ } train \text{ } in \text{ } 5 \text{ } min] = 0.607$$

and $P[at\ least\ 1\ train\ in\ next\ 4\ min] = 1 - P[0\ train\ in\ next\ 4\ min]$

$$= 1 - \frac{e^{-0.4}(0.4)^0}{0!} = 0.33$$

Hence, $P[0\ train\ in\ 5\ min\ \&\ at\ least\ 1\ train\ in\ next\ 4\ min]$
$$= P[0\ train\ in\ 5\ min] \times P[at\ least\ 1\ train\ in\ next\ 4\ min]$$
$$= 0.607 \times 0.33 = 0.2$$

Example 2.6

Pure Aloha is a packet radio network, originated at the University of Hawaii, that provides communication between a central computer and various remote data terminals (nodes). When a node has a packet to send, it will transmit it immediately. If the transmitted packet collides with other packets, the node concerned will re-transmit it after a random delay τ. Calculate the throughput of this pure Aloha system.

Solution

For simplicity, let us make the following assumptions:

(i)   The packet transmission time is one (one unit of measure).
(ii) The number of nodes is large, hence the total arrival of packets from all nodes is Poisson with rate λ.
(iii) The random delay τ is exponentially distributed with density function $\beta e^{-\beta \tau}$, where β is the node's retransmission attempt rate.

Given these assumptions, if there are n node waiting for the channel to re-transmit their packets, then the total packet arrival presented to the channel can be assumed to be Poisson with rate $(\lambda + n\beta)$ and the throughput S is then given by

$$S = (\lambda + n\beta)P[a\ successful\ transmission]$$
$$= (\lambda + n\beta)P_{succ}$$

From Figure 2.15, we see that there will be no packet collision if there is only one packet arrival within two units of time. Since the total arrival of packets is assumed to be Poisson, we have

$$P_{succ} = e^{-2(\lambda + n\beta)}$$

and hence

$$S = (\lambda + n\beta)e^{-2(\lambda + n\beta)}$$

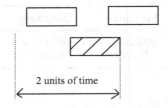

Figure 2.15 Vulnerable period of a transmission

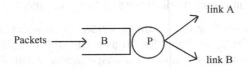

Figure 2.16 A schematic diagram of a switching node

Problems

1. Figure 2.16 shows a schematic diagram of a node in a packet switching network. Packets which are exponentially distributed arrive at the big buffer B according to a Poisson process. Processor P is a switching processor that takes a time τ, which is proportional to the packets' length, to route a packet to either of the two links A and B. We are interested in the average transition time that a packet takes to complete its transmission on either link, so how do you model this node as a queueing network?

2. Customers arrive at a queueing system according to a Poisson process with mean λ. However, a customer entering the service facility will visit the exponential server k times before he/she departs from the system. In each of the k visits, the customer receives an exponentially distributed amount of time with mean $1/k\mu$.

   (i) Find the probability density function of the service time.
   (ii) How do you describe the system in Kendall notation?

3. A communication line linking devices A and B is operating at 4800 bps. If device A sends a total of 30 000 characters of 8 bits each down the line in a peak minute, what is the resource utilization of the line during this minute?

4. All the telephones in the Nanyang Technological University are connected to the university central switchboard, which has 120 external lines to the local telephone exchange. The voice traffic generated by its employees in a typical working day is shown as:

	Incoming	Outgoing	Mean holding time
Local calls	500 calls/hr	480 calls/hr	2.5 min
Long-distance calls	30 calls/hr	10 calls/hr	1 min

Calculate the following:

(i)   the total traffic offered to the PABX.
(ii) the overall mean holding time of the incoming traffic.
(iii) the overall mean holding time of the outgoing traffic.

5. Consider a car-inspection centre where cars arrive at a rate of 1 every 30 seconds and wait for an average of 5 minutes (inclusive of inspection time) to receive their inspections. After the inspection, 20% of the car owners stay back and spend an average of 10 minutes in the centre's cafeteria. What is the average number of cars within the premise of the inspection centre (inclusive of the cafeteria)?

6. If packets arrive at a switching node according to a Poisson process with rate λ, show that the time interval X taken by the node to receive k packets is an Erlang-k random variable with parameters n and λ.

7. Jobs arrive at a single processor system according to a Poisson process with an average rate of 10 jobs per second. What is the probability that no jobs arrive in a 1-second period? What is the probability that 5 or fewer jobs arrive in a 1-second period? By letting t be an arbitrary point in time and T the elapsed time until the fifth job arrives after t, find the expected value and variance of T?

8. If $X_1, X_2, \ldots, X_n$ are independent exponential random variables with parameters $\lambda_1, \lambda_2, \ldots, \lambda_n$ respectively, show that the random variable $Y = \min\{X_1, X_2, \ldots X_n\}$ has an exponential distribution with parameter $\lambda = \lambda_1 + \lambda_2 + \ldots + \lambda_n$.

3

Discrete and Continuous Markov Processes

In Chapter 2, we derived Little's theorem which gives us a basic means to study performance measures of a queueing system. Unfortunately, if we take a closer look at this expression, the only quantity that we have prior knowledge of is probably the average arrival rate. The other two quantities are generally unknown and they are exactly what we want to determine for that particular queueing system.

To exploit the full potential of Little's theorem, we need other means to calculate either one of the two quantities. It turns out that it is easier to determine N as the number of customers in a queueing system that can be modelled as a continuous-time Markov process, a concept that we will study in this chapter. Once the probability mass function of the number of customers in a queueing system is obtained using the Markov chain, the long-term average N can then be easily computed.

Before we embark on the theory of Markov processes, let us look at the more general random processes – the so-called stochastic processes. The Markov process is a special class of stochastic processes that exhibits particular kinds of dependencies among the random variables within the same process. It provides the underlying theory of analysing queueing systems. In fact, each queueing system can, in principle, be mapped onto an instance of a Markov process or its variants (e.g. Imbedded Markov process) and mathematically analysed. We shall discuss this in detail later in the chapter.

3.1 STOCHASTIC PROCESSES

Simply put, a *stochastic process* is a mathematical model for describing an empirical process that changes with an index, which is usually the time in most of the real-life processes, according to some probabilistic forces. More specifically, a stochastic process is a family of random variables $\{X(t), t \in T\}$ defined on some probability space and indexed by a parameter $t\{t \in T\}$, where t is usually called the time parameter. The probability that $X(t)$ takes on a value, say i and that is $P[X(t) = i]$, is the range of that probability space.

In our daily life we encounter many stochastic processes. For example, the price $P_{st}(t)$ of a particular stock counter listed on the Singapore stock exchange as a function of time is a stochastic process. The fluctuations in $P_{st}(t)$ throughout the trading hours of the day can be deemed as being governed by probabilistic forces and hence a stochastic process. Another example will be the number of customers calling at a bank as a function of time.

Basically, there are three parameters that characterize a stochastic process:

(1) State space
The values assumed by a random variable $X(t)$ are called '*states*' and the collection of all possible values forms the '*state space*' of the process. If $X(t) = i$ then we say the process is in state i. In the stock counter example, the state space is the set of all prices of that particular counter throughout the day.

If the state space of a stochastic process is finite or at most countably infinite, it is called a '*discrete-state*' process, or commonly referred to as a stochastic chain. In this case, the state space is often assumed to be the non-negative integers $\{0, 1, 2, \ldots\}$. The stock counter example mentioned above is a discrete-state stochastic chain since the price fluctuates in steps of few cents or dollars.

On the other hand, if the state space contains a finite or infinite interval of the real numbers, then we have a '*continuous-state*' process. At this juncture, let us look at a few examples about the concept of 'countable infinite' without going into the mathematics of set theory. For example, the set of positive integer numbers $\{n\}$ in the interval $[a, b]$ is finite or countably infinite, whereas the set of real numbers in the same interval $[a, b]$ is infinite.

In the subsequent study of queueing theory, we are going to model the number of customers in a queueing system at a particular time as a Markov chain and the state represents the actual number of customers in the system. Hence we will restrict our discussion to the discrete-space stochastic processes.

(2) Index parameter
As mentioned above, the index is always taken to be the time parameter in the context of applied stochastic processes. Similar to the state space, if a process changes state at discrete or finite countable time instants, we have a

Table 3.1 Classifications of stochastic processes

Time Parameter	State Space	
	Discrete	Continuous
Discrete time	discrete-time stochastic chain	discrete-time stochastic process
Continuous time	continuous-time stochastic chain	continuous-time stochastic process

'*discrete (time) – parameter*' process. A discrete-parameter process is also called a *stochastic sequence*. In this case, we usually write $\{X_k \mid k \in N = (0, 1, 2, \ldots)\}$ instead of the enclosed time parameter $\{X(t)\}$. Using the stock price example again, if we are only interested in the closing price of that counter then we have a stochastic sequence.

On the other hand, if a process changes state (or in the terminology of Markov theory makes a '*transition*') at any instant on the time axis, then we have a '*continuous (time) – parameter*' process. For example, the number of arrivals of packets to a router during a certain time interval $[a, b]$ is a continuous-time stochastic chain because $t \in [a, b]$ is a continuum.

Table 3.1 gives the classification of stochastic processes according to their state space and time parameter.

(3) Statistical dependency

Statistical dependency of a stochastic process refers to the relationships between one random variable and others in the same family. It is the main feature that distinguishes one group of stochastic processes from another.

To study the statistical dependency of a stochastic process, it is necessary to look at the nth order joint (cumulative) probability distribution which describes the relationships among random variables in the same process. The nth order joint distribution of the stochastic process is defined as

$$F(\tilde{x}) = P[X(t_1) \leq x_1, \ldots, X(t_n) \leq x_n] \tag{3.1}$$

where

$$\tilde{x} = (x_1, x_2, \ldots, x_n)$$

Any realization of a stochastic process is called a sample path. For example, a sample path of tossing a coin n times is {head, tail, tail, head, . . . , head}.

Markov processes are stochastic processes which exhibit a particular kind of dependency among the random variables. For a Markov process, its future probabilistic development is dependent only on the most current state, and how the process arrives at the current position is irrelevant to the future concern. More will be said about this process later.

In the study of stochastic processes, we are generally interested in the probability that $X(t)$ takes on a value i at some future time t, that is $\{P[X(t) = i]\}$, because precise knowledge cannot be had about the state of the process in future times. We are also interested in the steady state probabilities if the probability converges.

Example 3.1

Let us denote the day-end closing price of a particular counter listed on the Singapore stock exchange on day k as X_k. If we observed the following closing prices from day k to day $k + 3$, then the following observed sequence $\{X_k\}$ is a stochastic sequence:

$$X_k = \$2.45 \quad X_{k+1} = \$2.38$$

$$X_{k+2} = \$2.29 \quad X_{k+3} = \$2.78$$

However, if we are interested in the fluctuations of prices during the trading hours and assume that we have observed the following prices at the instants $t_1 < t_2 < t_3 < t_4$, then the chain $\{X(t)\}$ is a continuous-time stochastic chain:

$$X(t_1) = \$2.38 \quad X(t_2) = \$2.39$$

$$X(t_3) = \$2.40 \quad X(t_4) = \$2.36$$

3.2   DISCRETE-TIME MARKOV CHAINS

The *discrete-time Markov chain* is easier to conceptualize and it will pave the way for our later introduction of the continuous Markov processes, which are excellent models for the number of customers in a queueing system.

As mentioned in Section 3.1, a Markov process is a stochastic process which exhibits a simple but very useful form of dependency among the random variables of the same family, namely the dependency that each random variable in the family has a distribution that depends only on the immediate preceding random variable. This particular type of dependency in a stochastic process was first defined and investigated by the Russian mathematician Andrei A Markov and hence the name Markov process, or Markov chain if the state space is discrete. In the following sections we will be merely dealing with only the discrete-state Markov processes; we will use Markov chain or process interchangeably without fear of confusion.

As an illustration to the idea of a Markov chain vs a stochastic process, let us look at a coin tossing experiment. Firstly, let us define two random variables; namely $X_k = 1$ (or 0) when the outcome of the kth trial is a 'head' (or tail), and Y_k = accumulated number of 'heads' so far. Assuming the system starts in state

Table 3.2 A sample sequence of Bernoulli trials

	H	H	T	T	H	T	H	H	H	T
X_k	1	1	0	0	1	0	1	1	1	0
Y_k	1	2	2	2	3	3	4	5	6	6

Zero ($Y_0 = 0$) and has the following sequence of the outcomes, as shown in Table 3.2.

then we see that X_k defines a chain of random variables or in other words a stochastic process, whereas Y_k forms a Markov chain as its values depend only on the cumulative outcomes and the preceding stage of the chain. That is

$$Y_k = Y_{k-1} + X_k \qquad (3.2)$$

The probabilities that there are, say, five accumulated 'heads' at any stage depends only on the number of 'heads' accumulated at the preceding stage (it must be four or five) together with the fixed probability X_k on a given toss.

3.2.1 Definition of Discrete-time Markov Chains

Mathematically, a stochastic sequence $\{X_k, k \in T\}$ is said to be a discrete-time Markov chain if the following conditional probability holds for all i, j and k:

$$P[X_{k+1} = j \mid X_0 = i_0, X_1 = i_1, \ldots, X_{k-1} = i_{k-1}, X_k = i] = P[X_{k+1} = j \mid X_k = i] \quad (3.3)$$

The above expression simply says that the $(k + 1)$th probability distribution conditional on all preceding ones equals the $(k + 1)$th probability distribution conditional on the kth; $k = 0, 1, 2, \ldots$. In other words, the future probabilistic development of the chain depends only on its current state (kth instant) and not on how the chain has arrived at the current state. The past history has been completely summarized in the specification of the current state and the system has no memory of the past – a *'memoryless'* chain. This *'memoryless'* characteristic is commonly known as the *Markovian* or *Markov property*.

The conditional probability at the right-hand side of Equation (3.3) is the probability of the chain going from state i at time step k to state j at time step $k + 1$ – the so-called *(one-step) transitional probability*. In general $P[X_{k+1} = j \mid X_k = i]$ is a function of time and in Equation (3.3) depends on the time step k. If the transitional probability does not vary with time, that is, it is invariant with respect to time epoch, then the chain is known as a *time-homogeneous Markov chain*. Using a short-hand notation, we write the conditional probability as

$$p_{ij} = P[X_{k+1} = j \mid X_k = i]$$

dropping the time index. Throughout this book, we will assume that all Markov processes that we deal with are time homogeneous.

For notational convention, we usually denote the state space of a Markov chain as $\{0, 1, 2, \ldots\}$. When $X_k = j$, the chain is said to be in state j at time k and we define the probability of finding the chain in this state using the new notation:

$$\pi_j^{(k)} \equiv P[X_k = j] \tag{3.4}$$

When a Markov chain moves from one state to another, we say the system makes a '*transition*'. The graphical representation of these dynamic changes of state, as shown in Figure 3.1, is known as the '*state transition diagram*' or '*transition diagram*' for short. In the diagram, the nodes represent states and the directed arcs between nodes represent the one-step transition probabilities. Those self-loops indicate the probabilities of remaining in the same state at the next time instant. In the case of a discrete-time Markov chain, the transitions between states can take place only at some integer time instants $0, 1, 2, \ldots k$, whereas transitions in the continuous case may take place at any instant of time.

As the system must transit to another state or remain in its present state at the next time step, we have

$$\sum_j p_{ij} = 1 \quad \& \quad 1 - p_{ii} = \sum_{j \neq i} p_{ij} \tag{3.5}$$

The conditional probability shown in Equation (3.3) expresses only the dynamism (or movement) of the chain. To characterize a Markov chain completely, it is necessary to specify the starting point of the chain, or in other words, the initial probability distribution $P[X_0 = i]$ of the chain. Starting with the initial state, it is in principle now possible to calculate the probabilities of finding the chain in a particular state at a future time using the total probability theorem:

$$P[X_{k+1} = j] = \sum_{i=0}^{\infty} P[X_{k+1} = j | X_k = i] P[X_k = i] = \sum_{i=0}^{\infty} \pi_i^{(k)} p_{ij} \tag{3.6}$$

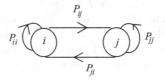

Figure 3.1   State transition diagram

The underlying principles of a Markov chain are best demonstrated by examples, so instead of dwelling on the mathematical formulation of Markov chains, we look at an example which illustrates the dynamics of a Markov chain and the use of Equation (3.6).

Example 3.2

A passenger lift in a shopping complex of three storeys is capable of stopping at every floor, depending on the passengers' traffic pattern. If the lift takes one time interval to go from one destination to another regardless of the number of floors between them, and the passenger traffic pattern is as shown in Table 3.3.

Then the position of the lift at the end of some time intervals in the future is clearly a Markov chain as the lift position at the next time step depends completely on its current position. Its state transition diagram is depicted in Figure 3.2.

Let X_k denotes the level at which we find the lift after k transitions and X_0 be the lift's initial position at time 0. $\pi_i^{(k)}$ is the probability of the lift in state i after k transitions.

We are given that the lift is at ground floor level at time 0. It is equivalent to saying that

Table 3.3 Passengers' traffic demand

Lift present position (current state)	Probability of going to the next level		
	ground floor (state 0)	1st floor (state 1)	2nd floor (state 2)
ground floor (state 0)	0	0.5	0.5
1st floor (state 1)	0.75	0	0.25
2nd floor (state 2)	0.75	0.25	0

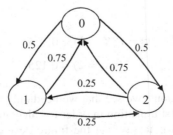

Figure 3.2 State transition diagram for the lift example

$$\pi_0^0 \equiv P(X_0 = 0) = 1$$
$$\pi_1^0 \equiv P(X_0 = 1) = 0$$
$$\pi_2^0 \equiv P(X_0 = 2) = 0$$

Or in vector form

$$\tilde{\pi}^{(0)} = [\pi_0^0, \pi_1^0, \pi_2^0] = [1, 0, 0]$$

(i) The probability of the lift's position after 1st transition:

$$\begin{aligned}
\pi_0^1 &= P(X_1 = 0) \\
&= \sum_i P(X_1 = 0 | X_0 = i) P(X_0 = i) \\
&= 1 \times 0 + 0 \times 0.5 + 0 \times 0.5 = 0
\end{aligned}$$

$$\begin{aligned}
\pi_1^1 &\equiv P(X_1 = 1) \\
&= \sum_i P(X_1 = 1 | X_0 = i) P(X_0 = i) \\
&= 0.5
\end{aligned}$$

Similarly

$$\pi_2^1 = P(X_1 = 2) = 0.5$$

(ii) The probability of the lift's position after 2nd transition:

$$\begin{aligned}
\pi_0^2 &\equiv P(X_2 = 0) \\
&= \sum_i P(X_2 = 0 | X_1 = i) P(X_1 = i) \\
&= 0 \times 0 + 0.75 \times 0.5 + 0.75 \times 0.5 \\
&= 0.75
\end{aligned}$$

and

$$\pi_1^2 = P(X_2 = 1) = 0.125$$
$$\pi_2^2 = P(X_2 = 2) = 0.125$$

From the above calculations, we see that we need only the probabilities of the lift's position after the 1st transition in order to calculate its probabilities after the 2nd transition. That is to say that the future development of the process depends only on its current position.

3.2.2 Matrix Formulation of State Probabilities

Students who have refreshed their memory of matrices in Chapter 1 will recognize that the above calculations can be formulated more succinctly in terms of matrix operations. Let us express the transition probabilities in an $n \times n$ square matrix $\tilde{P}$, assuming there are n states. The square matrix is known as *transition probability matrix*, or *transition matrix* for short:

$$\tilde{P} = (p_{ij}) = \begin{pmatrix} p_{11} & p_{12} & \cdots & p_{1n} \\ p_{21} & p_{22} & \cdots & p_{2n} \\ \cdots & & p_{ij} & \\ \cdots & & & p_{nn} \end{pmatrix} \tag{3.7}$$

The element p_{ij} is the transition probability defined in Equation (3.3). If the number of states is finite, say n, then we have a $n \times n$ matrix $\tilde{P}$, otherwise the matrix is infinite. Since the probability of going to all other states, including itself, should sum to unity, as shown in Equation (3.5), the sum of each row in the above matrix should equal to unity, that is:

$$\sum_j p_{ij} = 1$$

A matrix with each row sums to unity and all elements positive or zero are called a *stochastic matrix*. A Markov chain is completely characterized by this (one-step) transition probability matrix together with the initial probability vector.

Similarly, we express the state probabilities at each time interval as a row vector:

$$\tilde{\pi}^{(k)} = (\pi_0^{(k)}, \pi_1^{(k)}, \ldots, \pi_n^{(k)}) \tag{3.8}$$

Using these matrix notations, the calculations shown in Example 3.2 can be formulated as

$$\tilde{\pi}^{(1)} = \tilde{\pi}^{(0)} \tilde{P}$$
$$\tilde{\pi}^{(2)} = \tilde{\pi}^{(1)} \tilde{P}$$
$$\cdots$$
$$\tilde{\pi}^{(k)} = \tilde{\pi}^{(k-1)} \tilde{P} \tag{3.9}$$

Back substituting the $\tilde{\pi}^{(i)}$, we have from Equation (3.9) the following equation:

$$\tilde{\pi}^{(k)} = \tilde{\pi}^{(0)} \tilde{P}^{(k)} \tag{3.10}$$

where

$$\tilde{P}^{(k)} = \tilde{P} \cdot \tilde{P}^{(k-1)} = \tilde{P}^k \tag{3.11}$$

$\tilde{P}^k$, the so-called *k-step* transition matrix, is the k-fold multiplication of the one-step transition matrix by itself. We define $\tilde{P}^{(0)} = I$.

Equations (3.10) and (3.11) give us a general method for calculating the state probability k steps into a chain. From matrix operations, we know that

$$\tilde{P}^{(k+l)} = \tilde{P}^{(k)} \times \tilde{P}^{(l)} \tag{3.12}$$

or

$$P_{ij}^{k+l} = \sum_{k=0}^{n} P_{ik}^k P_{kj}^l \tag{3.13}$$

These two equations are the well-known Chapman–Kolmogorov forward equations.

Example 3.3

For the lift example, the transition probability matrix is

$$\tilde{P} = \begin{pmatrix} 0 & 0.5 & 0.5 \\ 0.75 & 0 & 0.25 \\ 0.75 & 0.25 & 0 \end{pmatrix}$$

The transition matrices for the first few transitions can be computed as

$$\tilde{P}^{(2)} = \tilde{P}^{(1)} \times \tilde{P}^{(1)} = \begin{pmatrix} 0.75 & 0.125 & 0.125 \\ 0.1875 & 0.4375 & 0.375 \\ 0.1875 & 0.375 & 0.4375 \end{pmatrix}$$

$$\tilde{P}^{(3)} = \begin{pmatrix} 0.1875 & 0.4063 & 0.4062 \\ 0.6094 & 0.1875 & 0.2031 \\ 0.6094 & 0.2031 & 0.1875 \end{pmatrix}$$

$$\tilde{P}^{(4)} = \begin{pmatrix} 0.6094 & 0.1953 & 0.1953 \\ 0.21930 & 0.3555 & 0.3515 \\ 0.2930 & 0.3515 & 0.3554 \end{pmatrix}$$

If the lift is in state 0 at time 0, i.e. $\tilde{\pi}^{(0)} = (1, 0, 0)$, then we have

$$\tilde{\pi}^{(1)} = \tilde{\pi}^{(0)} \times \tilde{P}^{(1)} = (0, 0.5, 0.5)$$

$$\tilde{\pi}^{(2)} = (0.75, 0.125, 0.125)$$

$$\tilde{\pi}^{(3)} = (0.1875, 0.4062, 0.4063)$$

If $\tilde{\pi}^{(0)} = (0, 1, 0)$, we have

$$\tilde{\pi}^{(1)} = \tilde{\pi}^{(0)} \times \tilde{P}^{(1)} = (0.75, 0, 0.25)$$

$$\tilde{\pi}^{(2)} = (0.1875, 0.4375, 0.375)$$

$$\tilde{\pi}^{(3)} = (0.6095, 0.1875, 0.2031)$$

If $\tilde{\pi}^{(0)} = (0, 0, 1)$, we have

$$\tilde{\pi}^{(1)} = \tilde{\pi}^{(0)} \times \tilde{P}^{(1)} = (0.75, 0.25, 0)$$

$$\tilde{\pi}^{(2)} = (0.1875, 0.3750, 0.4375)$$

$$\tilde{\pi}^{(3)} = (0.6094, 0.2031, 0.1875)$$

3.2.3   General Transient Solutions for State Probabilities

Students who have gone through Example 3.3 will be quick to realize that we were actually calculating the transient response of the Markov chain, which is the probability of finding the lift position at each time step. In this section, we present an elegant z-transform approach of finding a general expression for $\tilde{\pi}^{(k)}$.

First we define a matrix z-transform of $\tilde{\pi}^{(k)}$ as

$$\tilde{\pi}(z) \equiv \sum_{k=0}^{\infty} \tilde{\pi}^{(k)} z^k \tag{3.14}$$

Multiplying the expression (3.9) by z^k and summing it from $k = 1$ to infinity, we have

$$\sum_{k=1}^{\infty} \tilde{\pi}^{(k)} z^k = \sum_{k=1}^{\infty} \tilde{\pi}^{(k-1)} \tilde{P} z^k$$

The left-hand side is simply the z-transform of $\tilde{\pi}^{(k)}$ without the first term. On the right-hand side, $\tilde{P}$ is just a scalar constant. Adjusting the index, we have

$$\tilde{\pi}(z) - \tilde{\pi}^{(0)} = z\tilde{P}\sum_{k=1}^{\infty}\tilde{\pi}^{(k-1)}z^{k-1}$$

$$\tilde{\pi}(z) = \tilde{\pi}^{(0)}(\tilde{I} - z\tilde{P})^{-1} \tag{3.15}$$

where $\tilde{I}$ is the identity matrix and the superscript (-1) denotes the matrix inverse. If the inverse exists, we could obtain the transient solution for the state probability by carrying out the inverse z-transform of Equation (3.15). Comparing Equations (3.15) and (3.10), we see that the k-step transition matrix is given by

$$\tilde{P}^k = (\tilde{I} - z\tilde{P})^{-1} \tag{3.16}$$

Coupling with the initial state probability vector, we can then calculate the state probability vector k-step into the future.

Besides the z-transform method, there is another way of finding $\tilde{P}^{(k)}$. Recall from the section on 'Eigenvalues, Eigenvectors and Spectral Representation' in Chapter 1, if the eigenvalues λ_i of $\tilde{P}$ are all distinct, then $\tilde{P}^{(k)}$ can be expressed as

$$\begin{aligned}\tilde{P}^{(k)} &= N\Lambda^k N^{-1}\\ &= \lambda_1^k B_1 + \lambda_2^k B_2 \ldots + \lambda_n^k B_n\end{aligned} \tag{3.17}$$

where N is a matrix that makes up of the eigenvectors of $\tilde{P}$, and B_i are matrices defined in that section. Students should not mistakenly think that this is an easier way of computing $\tilde{P}^{(k)}$. It is in general quite difficult to compute the eigenvalues and eigenvectors of a matrix.

Example 3.4

Returning to the lift example, let us find the explicit solution for $\tilde{\pi}^{(k)}$. First we find the matrix $(\tilde{I} - z\tilde{P})$:

$$(\tilde{I} - z\tilde{P}) = \begin{pmatrix} 1, & -\dfrac{1}{2}z, & -\dfrac{1}{2}z \\[2mm] -\dfrac{3}{4}z, & 1, & -\dfrac{1}{4}z, \\[2mm] -\dfrac{3}{4}z, & -\dfrac{1}{4}z, & 1 \end{pmatrix}$$

To find the inverse, we need to find the determinant of the matrix:

$$\Delta \equiv \det(\tilde{I} - z\tilde{P}):$$

$$\Delta \equiv \det(\tilde{I} - z\tilde{P}) = \det \begin{pmatrix} 1, & -\dfrac{1}{2}z, & -\dfrac{1}{2}z \\ -\dfrac{3}{4}z, & 1, & -\dfrac{1}{4}z, \\ -\dfrac{3}{4}z, & -\dfrac{1}{4}z, & 1 \end{pmatrix} = (1-z)\left(1+\dfrac{1}{4}z\right)\left(1+\dfrac{3}{4}z\right)$$

and the inverse of $(\tilde{I} - z\tilde{P})$ is given by

$$(\tilde{I} - z\tilde{P})^{-1} = \dfrac{1}{\Delta} \begin{pmatrix} 1-\dfrac{1}{16}z^2, & \dfrac{1}{2}z+\dfrac{1}{8}z^2, & \dfrac{1}{2}z+\dfrac{1}{8}z^2 \\ \dfrac{3}{4}z+\dfrac{3}{16}z^2, & 1-\dfrac{3}{8}z^2, & \dfrac{1}{4}z+\dfrac{3}{8}z^2 \\ \dfrac{3}{4}z+\dfrac{3}{16}z^2, & \dfrac{1}{4}z+\dfrac{3}{8}z^2, & 1-\dfrac{3}{8}z^2 \end{pmatrix}$$

Carrying out partial fraction and grouping the result into three matrices, we obtain

$$(\tilde{I} - z\tilde{P})^{-1} = \dfrac{\dfrac{1}{7}}{1-z}\begin{pmatrix} 3 & 2 & 2 \\ 3 & 2 & 2 \\ 3 & 2 & 2 \end{pmatrix} + \dfrac{1}{1+\dfrac{1}{4}z}\begin{pmatrix} 0 & 0 & 0 \\ 0 & \dfrac{1}{2} & \dfrac{1}{2} \\ 0 & \dfrac{1}{2} & \dfrac{1}{2} \end{pmatrix} + \dfrac{\dfrac{1}{7}}{1+\dfrac{3}{4}z}\begin{pmatrix} 4 & -2 & -2 \\ -3 & \dfrac{3}{2} & \dfrac{3}{2} \\ -3 & \dfrac{3}{2} & \dfrac{3}{2} \end{pmatrix}$$

$$\tilde{P}^{(k)} = \dfrac{1}{7}\begin{pmatrix} 3 & 2 & 2 \\ 3 & 2 & 2 \\ 3 & 2 & 2 \end{pmatrix} + \left(-\dfrac{1}{4}\right)^k\begin{pmatrix} 0 & 0 & 0 \\ 0 & \dfrac{1}{2} & \dfrac{1}{2} \\ 0 & \dfrac{1}{2} & \dfrac{1}{2} \end{pmatrix} + \dfrac{1}{7}\left(-\dfrac{3}{4}\right)^k\begin{pmatrix} 4 & -2 & -2 \\ -3 & \dfrac{3}{2} & \dfrac{3}{2} \\ -3 & \dfrac{3}{2} & \dfrac{3}{2} \end{pmatrix}.$$

Once the k-step transition matrix is obtained, we can then calculate the transient solution using $\tilde{\pi}^{(k)} = \tilde{\pi}^{(0)}\tilde{P}^{(k)}$.

Example 3.5

Now, let us use the spectrum representation method to find the explicit solution for $P^{(k)}$. First we have to find the eigenvalues and eigenvectors of the transition probability matrix:

$$\tilde{P} = \begin{pmatrix} 0 & 0.5 & 0.5 \\ 0.75 & 0 & 0.25 \\ 0.75 & 0.25 & 0 \end{pmatrix}$$

The eigenvalues can be found by solving the equation $\det(\tilde{P} - \lambda \tilde{I}) = 0$. Forming the necessary matrix and taking the determinant, we have

$$(\lambda - 1)(4\lambda + 3)(4\lambda + 1) = 0$$

Now it remains for us to find the corresponding eigenvectors. For $\lambda = 1$, the corresponding eigenvector (f_1) can be found by solving the following set of simultaneous equations:

$$\begin{pmatrix} 0 & 0.5 & 0.5 \\ 0.75 & 0 & 0.25 \\ 0.75 & 0.25 & 0 \end{pmatrix} \begin{pmatrix} x_1 \\ x_2 \\ x_3 \end{pmatrix} = (1) \begin{pmatrix} x_1 \\ x_2 \\ x_3 \end{pmatrix}$$

Similarly, the other eigenvectors $(f_2$ and $f_3)$ can be found using the same method and we have

$$\tilde{N} = \begin{pmatrix} 1 & 1/3 & 0 \\ 1 & -1/4 & 1 \\ 1 & -1/4 & -1 \end{pmatrix} \quad \text{and} \quad \tilde{\Lambda} = \begin{pmatrix} 1 & 0 & 0 \\ 0 & -3/4 & 0 \\ 0 & 0 & -1/4 \end{pmatrix}$$

Finding the inverse of N, we have

$$\tilde{N}^{-1} = \begin{pmatrix} 3/7 & 2/7 & 2/7 \\ 12/7 & -6/7 & -6/7 \\ 0 & 1/2 & -1/2 \end{pmatrix}$$

Having obtained these matrices, we are ready to form the following matrices:

$$\tilde{B}_1 = \tilde{x}_1 \times \tilde{\pi}_1 = \begin{pmatrix} 1 \\ 1 \\ 1 \end{pmatrix} (3/7 \quad 2/7 \quad 2/7) = \begin{pmatrix} 3/7 & 2/7 & 2/7 \\ 3/7 & 2/7 & 2/7 \\ 3/7 & 2/7 & 2/7 \end{pmatrix}$$

$$\tilde{B}_3 = \tilde{x}_3 \times \tilde{\pi}_3 = \begin{pmatrix} 1/3 \\ -1/4 \\ -1/4 \end{pmatrix} (12/7 \quad -6/7 \quad -6/7) = \frac{1}{7} \begin{pmatrix} 4 & -2 & -2 \\ -3 & 3/2 & 3/2 \\ -3 & 3/2 & 3/2 \end{pmatrix} = \frac{1}{7} \begin{pmatrix} 4 & -2 & -2 \\ -3 & 3/2 & 3/2 \\ -3 & 3/2 & 3/2 \end{pmatrix}$$

$$\tilde{B}_2 = \tilde{x}_2 \times \tilde{\pi}_2 = \begin{pmatrix} 0 \\ 1 \\ -1 \end{pmatrix} (0 \quad 1/2 \quad -1/2) = \begin{pmatrix} 0 & 0 & 0 \\ 0 & 1/2 & 1/2 \\ 0 & 1/2 & 1/2 \end{pmatrix}$$

Therefore, the transient solution for the state probabilities is given by

$$\tilde{P}^{(k)} = (1)^k \tilde{B}_1 + (-1/4)^k \tilde{B}_2 + (-3/4)^k \tilde{B}_3$$

which is the same as that obtained in Example 3.4.

Example 3.6

Continuing with Example 3.4, we see that the *k-step* transition matrix is made up of three matrices; the first is independent of k. If we let $k \to \infty$, we have

$$\lim_{k \to \infty} \tilde{P}^{(k)} = \frac{1}{7} \begin{pmatrix} 3 & 2 & 2 \\ 3 & 2 & 2 \\ 3 & 2 & 2 \end{pmatrix} = \tilde{P}$$

Using expression

$$\lim_{k \to \infty} \tilde{\pi}^{(k)} = \lim_{k \to \infty} \tilde{\pi}^{(0)} \tilde{P}^{(k)}$$

we have

$$\lim_{k \to \infty} \tilde{\pi}^{(k)} = (1 \quad 0 \quad 0) \frac{1}{7} \begin{pmatrix} 3 & 2 & 2 \\ 3 & 2 & 2 \\ 3 & 2 & 2 \end{pmatrix} = \frac{1}{7}(3 \quad 2 \quad 2)$$

$$\lim_{k \to \infty} \tilde{\pi}^{(k)} = (0 \quad 1 \quad 0) \frac{1}{7} \begin{pmatrix} 3 & 2 & 2 \\ 3 & 2 & 2 \\ 3 & 2 & 2 \end{pmatrix} = \frac{1}{7}(3 \quad 2 \quad 2)$$

$$\lim_{k \to \infty} \tilde{\pi}^{(k)} = (0 \quad 0 \quad 1) \frac{1}{7} \begin{pmatrix} 3 & 2 & 2 \\ 3 & 2 & 2 \\ 3 & 2 & 2 \end{pmatrix} = \frac{1}{7}(3 \quad 2 \quad 2)$$

We see that we always have the same final state probability vector when $k \to \infty$, regardless of the initial state probability if $\tilde{\pi} = \lim_{k \to \infty} \tilde{\pi}^{(k)}$ exists.

3.2.4 Steady-state Behaviour of a Markov Chain

From Example 3.6, we see that the limiting value of $\tilde{P} = \lim_{k \to \infty} \tilde{P}^{(k)}$ is indepe-dent of k and there exists a limiting value of the state probabilities $\tilde{\pi} = \lim_{k \to \infty} \tilde{\pi}^{(k)}$, which is independent of the initial probability vector.

We say these chains have *steady states* and $\tilde{\pi}$ is called the *stationary prob-ability* or *stationary distribution*. The stationary probability can be calculated from Equation (3.9) as

$$\lim_{k \to \infty} \tilde{\pi}^{(k)} = \lim_{k \to \infty} \tilde{\pi}^{(k-1)} \tilde{P} \to \tilde{\pi} = \tilde{\pi} \tilde{P} \tag{3.18}$$

For most of the physical phenomena and under very general conditions that are of interest to us, limiting values always exist. We will state a theorem without proof but clarify some of the terms used in the theorem subsequently.

Theorem 3.1

A discrete Markov chain $\{X_k\}$ that is irreducible, aperiodic and time-homoge-neous is said to be ergodic. For an ergodic Markov chain, the limiting probabilities:

$$\pi_j = \lim_{k \to \infty} \pi_j^{(k)} = \lim_{k \to \infty} P[X_k = j] \quad j = 0, 1, \ldots$$

$$\tilde{\pi} = \lim_{k \to \infty} \pi^{(k)} \quad \text{(matrix notation)}$$

always exist and are independent of the initial state probability distribution. The stationary probabilities π_j are uniquely determined through the following equations:

$$\sum_j \pi_j = 1 \tag{3.19}$$

$$\pi_j = \sum_i \pi_i P_{ij} \tag{3.20}$$

Expressions (3.19) and (3.20) can be formulated using matrix operations as follows:

$$\tilde{\pi} \cdot \tilde{e} = 1$$

$$\tilde{\pi} = \tilde{\pi} \cdot \tilde{P}$$

where $\tilde{e}$ is a $1 \times n$ row vector with all entries equal to one.

To solve these two matrix equations, we first define a $n \times n$ matrix $\tilde{U}$ with all entries equal to one, then the first equation can be rewritten as $\tilde{\pi} \cdot \tilde{U} = \tilde{e}$. Students should note that these two matrices, $\tilde{U}$ or $\tilde{e}$, have the summation property. Any matrix, say $\tilde{A} = [a_{ij}]$, multiplied by $\tilde{U}$ or $\tilde{e}$ will give rise to a matrix in which every entry in a row is the same and is the sum of the corresponding row in $\tilde{A}$. For example:

$$\begin{pmatrix} a_{11} & a_{12} \\ a_{21} & a_{22} \end{pmatrix}\begin{pmatrix} 1 & 1 \\ 1 & 1 \end{pmatrix} = \begin{pmatrix} a_{11} + a_{12} & a_{11} + a_{12} \\ a_{21} + a_{22} & a_{21} + a_{22} \end{pmatrix}$$

Adding the two equations together, we have

$$\tilde{\pi}(\tilde{P} + \tilde{U} - \tilde{I}) = \tilde{e} \tag{3.21}$$

We will postpone our discussion of some terms, for example, aperiodic and irreducible, used in the theorem to the next section and instead look at how we could use this theorem to calculate the steady-state probabilities of a Markov chain.

Example 3.7

Continuing with the lift example, the transition probabilities are given by

$$\tilde{P} = (p_{ij}) = \begin{pmatrix} 0 & 1/2 & 1/2 \\ 3/4 & 0 & 1/4 \\ 3/4 & 1/4 & 0 \end{pmatrix}$$

and steady state probabilities $\tilde{\pi} = [\pi_0, \pi_1, \pi_2]$. Using Equation (3.18) and expanding it, we have

$$\pi_0 = 0\pi_0 + \frac{3}{4}\pi_1 + \frac{3}{4}\pi_2$$

$$\pi_1 = \frac{1}{2}\pi_0 + 0\pi_1 + \frac{1}{4}\pi_2$$

$$\pi_2 = \frac{1}{2}\pi_0 + \frac{1}{4}\pi_1 + 0\pi_2$$

This set of equations is not unique as one of the equations is a linear combination of others. To obtain a unique solution, we need to replace any one of them by the normalization equation:

$$\sum_i \pi_i = 1 \quad \text{i.e.} \quad \pi_0 + \pi_1 + \pi_2 = 1$$

Solving them, we have

$$\pi_0 = \frac{3}{7}, \quad \pi_1 = \frac{2}{7}, \quad \pi_2 = \frac{2}{7}$$

3.2.5 Reducibility and Periodicity of a Markov Chain

In this section, we clarify some of the terms used in Theorem 3.1. A Markov chain is said to be reducible if it contains more than one isolated closed set of states. For example, the following transition matrix has two isolated closed sets of states, as shown in Figure 3.3. Depending on which state the chain begins with, it will stay within one of the isolated closed sets and never enter the other closed set.

$$\tilde{P} = \begin{pmatrix} \frac{1}{2} & \frac{1}{2} & 0 & 0 \\ \frac{1}{2} & \frac{1}{2} & 0 & 0 \\ 0 & 0 & \frac{1}{2} & \frac{1}{2} \\ 0 & 0 & \frac{1}{2} & \frac{1}{2} \end{pmatrix}$$

An irreducible Markov chain is one which has only one closed set and all states in the chain can be reached from any other state. Any state j is said to be reachable from any other states i if it is possible to go from state i to state j in a finite number of steps according to the given transition probability matrix. In other words, if there exists a k where $\infty > k \geq 1$ such that $p_{ij}^{(k)} > 0$, $\forall i, j$.

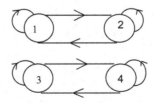

Figure 3.3 Transition diagram of two disjoined chains

A Markov chain is said to be periodic with period τ if it only returns to a particular state (say i) after $n\tau(n = 1, 2, \ldots)$ steps. Otherwise, it is aperiodic. In an irreducible Markov chain, all states are either all aperiodic or all periodic with the same period.

Example 3.8

The closed set of states in a probability transition matrix can be difficult to recognize. The following probability transition matrix represents the transitions shown in Figure 3.3, except now nodes 1 and 3 form a closed set and 2 and 4 another (Figure 3.4).

$$\tilde{P} = \begin{pmatrix} \frac{1}{2} & 0 & \frac{1}{2} & 0 \\ 0 & \frac{1}{2} & 0 & \frac{1}{2} \\ \frac{1}{2} & 0 & \frac{1}{2} & 0 \\ 0 & \frac{1}{2} & 0 & \frac{1}{2} \end{pmatrix}$$

In fact, students can verify that the above given transition-rate matrix can be reduced to

$$\tilde{P} = \begin{pmatrix} \frac{1}{2} & \frac{1}{2} & 0 & 0 \\ \frac{1}{2} & \frac{1}{2} & 0 & 0 \\ 0 & 0 & \frac{1}{2} & \frac{1}{2} \\ 0 & 0 & \frac{1}{2} & \frac{1}{2} \end{pmatrix}$$

through some appropriate row and column interchanges.

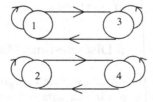

Figure 3.4 Transition diagram of two disjoined chains

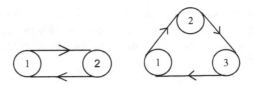

Figure 3.5   Periodic Markov chain

Example 3.9

The two simplest examples of periodic Markov chain are shown below. If we assume that the initial position of these two chains is state 1, then the first chain will always go back to state 1 after two transitions and the second chain after three transitions (Figure 3.5).

The transition probability matrices for these two Markov chains are

$$\tilde{P}_1 \begin{pmatrix} 0 & 1 \\ 1 & 0 \end{pmatrix} \quad \text{and} \quad \tilde{P}_2 = \begin{pmatrix} 0 & 1 & 0 \\ 0 & 0 & 1 \\ 1 & 0 & 0 \end{pmatrix}$$

respectively. If we compute the k-step transition probability matrices, we have

$$\tilde{P}_1 = \tilde{P}_1^{(3)} = P_1^{(5)} = \ldots = \begin{pmatrix} 0 & 1 \\ 1 & 0 \end{pmatrix}$$

$$\tilde{P}_2 = \tilde{P}_2^{(4)} = \tilde{P}_2^{(7)} = \ldots = \begin{pmatrix} 0 & 1 & 0 \\ 0 & 0 & 1 \\ 1 & 0 & 0 \end{pmatrix}$$

We see that these two Markov chains have periods 2 and 3, respectively. Note that these two chains still have a stationary probability distribution though they are periodic in nature. Students can verify that the stationary probabilities are given as

$$\tilde{\pi} = [0.5 \ 0.5] \quad \textit{two-state chain}$$

$$\tilde{\pi} = [1/3 \ \ 1/3 \ \ 1/3] \quad \textit{three-state chain}$$

3.2.6   Sojourn Times of a Discrete-time Markov Chain

Sojourn time of a discrete-time Markov chain in a given state, say i, refers to the number of time units it spends in that state. We will restrict our discussion to the homogeneous Markov chains here.

Assuming the Markov chain has just entered state i, the probability of it staying in this state at the next time step is given by $p_{ij} = 1 - \sum_{i \neq j} p_{ij}$. If it stays in this state, again the probability of it staying in that state at the next time step is also p_{ii}. Hence, we see that the time a Markov chain stays in a state i, sojourn time R_i, is a geometrically distributed random variable with *pmf*:

$$P[R_i = k] = (1 - p_{ii}) p_{ii}^{k-1}, \quad \forall k = 1, 2, 3, \ldots \tag{3.22}$$

And the expected sojourn time is given by

$$E[R_i] = \frac{1}{1 - p_{ii}}$$

The previous result of the sojourn time is in line with the memoryless property of a Markov chain. The geometric distribution is the only discrete probability distribution that possesses such a memoryless property. Exponential distribution is the continuous version that also possesses such a property and we will show later that the sojourn times of a continuous-time Markov chain is exponentially distributed.

3.3   CONTINUOUS-TIME MARKOV CHAINS

Having discussed the discrete-time Markov chain, we are now ready to look at its continuous-time counterpart. Conceptually, there is no difference between these two classes of Markov chain; the past history of the chain is still being summarized in its present state and its future development can be inferred from there. In fact, we can think of the continuous-time Markov chain as being the limiting case of the discrete type.

However, there is a difference in mathematical formulation. In a continuous Markov chain, since transitions can take place at any instant of a time continuum, it is necessary now to specify how long a process has stayed in a particular state before a transition take place.

In some literature, the term 'Markov process' is also used to refer to a continuous-time Markov chain. We use that term occasionally in subsequent chapters.

3.3.1   Definition of Continuous-time Markov Chains

The definition of a continuous-time Markov chain parallels that of a discrete-time Markov chain conceptually. It is a stochastic process $\{X(t)\}$ in which the future probabilistic development of the process depends only on its present state and not on its past history. Mathematically, the process should satisfy the following conditional probability relationships for $t_1 < t_2 < \ldots < t_k$:

$$P[X(t_{k+1}) = j | X(t_1) = i_1, X(t_2) = t_2, \ldots X(t_k) = i_k] = P[X(t_{k+1}) = j | X(t_k) = i_k]$$

(3.23)

From the early discussion, we know that in the treatment of continuous-time Markov chains we need to specify a transition scheme by which the system goes from one state to another. The transition probability alone is not sufficient to completely characterize the process. Instead of dwelling on its formal theoretical basics which can be quite involved, we will take the following definition as the point of departure for our subsequent discussions and develop the necessary probability distribution. And again, we will focus our attention on the homogeneous case.

Definition 3.1

For a continuous Markov chain which is currently in state i, the probability that the chain will leave state i and go to state $j(j \neq i)$ in the next infinitesimal amount of time Δt, no matter how long the chain has been in state i, is

$$p_{ij}(t, t + \Delta t) = q_{ij}\Delta t$$

where q_{ij} is the *instantaneous transition rate* of leaving state i for state j. In general, q_{ij} is a function of t. To simply the discussion, we assume that q_{ij} is independent of time; in other words, we are dealing with a homogeneous Markov chain. The total instantaneous transition rate at which the chain leaves state i is, therefore, $\sum_{j \neq i} q_{ij}$.

3.3.2   Sojourn Times of a Continuous-time Markov Chains

Analogous to the discrete-time Markov chain, the sojourn times of a continuous-time Markov chain is the time the chain spends in a particular state. From the earlier discussion of Markov chains, we know that the future probabilistic development of a chain is only related to the past history through its current position. Thus, the sojourn times of a Markov chain must be 'memoryless' and are exponentially distributed, as exponential distribution is the only continuous probability distribution that exhibits such a memoryless property.

We shall demonstrate that the sojourn times of a continuous-time Markov chain are indeed exponentially distributed using the above definition. We assume the chain has just entered state i, and will remain in that during an interval $[0, t]$.

Let τ_i be the random variable that denotes the time spent in state i. If we divide the time t into k equal intervals of Δt, such that $k\Delta t = t$, then for the sojourn time τ_i to be greater than t, there should not be any transition in any of these Δt intervals. From the above definition, we know that the probability of not having a transition in a time interval Δt is

$$\left(1 - \sum_{j \neq i} q_{ij} \Delta t\right)$$

Therefore, the probability that the sojourn is greater than t

$$P[\tau_i > t] = \lim_{k \to \infty} \left[1 - \sum_{j \neq i} q_{ij} \Delta t\right]^k$$

$$= \lim_{k \to \infty} \left[1 - \sum_{j \neq i} q_{ij} \frac{t}{k}\right]^k$$

$$= e^{-q_i t} \tag{3.24}$$

where $q_i = -\sum_{j \neq i} q_{ij}$. Hence, the sojourn time between transitions is given by

$$P[\tau_i \leq t] = 1 - P[\tau_i > t] = 1 - e^{-q_i t} \tag{3.25}$$

which is an exponential distribution.

3.3.3 State Probability Distribution

We will now turn our attention to the probability of finding the chain in a particular state – state probability. As usual, we define $\pi_j(t) = P[X(t) = j]$ and consider the probability change in an infinitesimal amount of time Δt:

$$\pi_j(t + \Delta t) = \sum_{i \neq j} \pi_i(t) q_{ij} \Delta t + \pi_j(t) \left[1 - \sum_{k \neq j} q_{jk} \Delta t\right] \tag{3.26}$$

The first term on the right is the probability that the chain is in state i at time t and makes a transition to state j in Δt. The second term is the probability that the chain is in state j and does not make a transition to any other states in Δt. Rearranging terms, dividing Equation (3.20) by Δt and taking limits, we have

$$\frac{d}{dt} \pi_j(t) = \sum_{i \neq j} \pi_i q_{ij} - \pi_j \sum_{k \neq j} q_{jk} \tag{3.27}$$

If we define $q_{jj} = -\sum\limits_{k \neq j} q_{ik}$ then the above expression can be re-written as

$$\frac{d}{dt}\pi_j(t) = \sum_i \pi_i(t)q_{ij} \qquad (3.28)$$

Let us define the following three matrices:

$$\tilde{\pi}(t) = (\pi_1(t), \pi_2(t), \ldots) \qquad (3.29)$$

$$\frac{d}{dt}\tilde{\pi}(t) = \left(\frac{d}{dt}\pi_1(t), \frac{d}{dt}\pi_2(t), \ldots\right) \qquad (3.30)$$

$$\tilde{Q} = (q_{ij}) = \begin{pmatrix} -\sum\limits_{j \neq 1} q_{1j} & q_{12} & q_{13} & \cdots \\ q_{21} & -\sum\limits_{j \neq 2} q_{2j} & q_{23} & \\ \cdots & & & \cdots \\ q_{n1} & \cdots & & -\sum\limits_{j \neq n} q_{nj} \end{pmatrix} \qquad (3.31)$$

We can then re-write Equation (3.27) or (3.28) in matrix form as

$$\frac{d}{dt}\tilde{\pi}(t) = \tilde{\pi}(t)\tilde{Q} \qquad (3.32)$$

The matrix $\tilde{Q}$ is known as the *infinitesimal generator* or *transition-rate matrix* as its elements are the instantaneous rates of leaving a state for another state. Recall from the section on 'Eigenvalues, Eigenvectors and Spectral Representation' in Chapter 1, one of the general solutions for the above matrix equation is given by

$$\tilde{\pi}(t) = e^{\tilde{Q}t} = \tilde{I} + \sum_{k=1}^{\infty} \frac{(\tilde{Q})^k t^k}{k!}$$

Similar to the discrete-time Markov chains, the limiting value of the state probability $\tilde{\pi} = \lim\limits_{t \to \infty} \tilde{\pi}(t)$ exists for an irreducible homogeneous continuous-time Markov chain and is independent of the initial state of the chain. That implies that $\frac{d}{dt}\tilde{\pi}(t) = 0$ for these limiting values. Setting the differential of the state probabilities to zero and taking limit, we have

$$0 = \tilde{\pi}\tilde{Q} \qquad (3.33)$$

where $\tilde{\pi} = (\pi_1, \pi_2, \ldots)$ and $\tilde{Q}$ is the transition-rates matrix defined in Equation (3.31).

We can see the distinct similarity in structure if we compare this equation with that for a discrete-time Markov chain, i.e. $\tilde{\pi} = \tilde{\pi}\tilde{P}$. These two equations uniquely describe the 'motion' of a Markov chain and are called stationary equations. Their solutions are known as the stationary distributions.

3.3.4 Comparison of Transition-rate and Transition-probability Matrices

Comparing the transition rate matrix (3.31) defined in Section 3.3.3 with the transition probability matrix (3.7) of Section 3.2.2, we see some similarities as well as distinctions between them.

First of all, each of these two matrices completely characterizes a Markov chain, i.e. $\tilde{Q}$ for a continuous Markov chain and $\tilde{P}$ for the discrete counterpart. All the transient and steady-state probabilities of the corresponding chain can in principle be calculated from them in conjunction with the initial probability vector.

The main distinction between them lies in the fact that all entries in $\tilde{Q}$ are transition rates whereas that of $\tilde{P}$ are probabilities. To obtain probabilities in $\tilde{Q}$, each entry needs to be multiplied by Δt, i.e. $q_{ij}\Delta t$.

Secondly, each row of the $\tilde{Q}$ matrix sums to zero instead of one, as in the case of $\tilde{P}$ matrix. The instantaneous transition rate of going back to itself is not defined in the continuous case. It is taken to be $q_{jj} = -\sum_{k \neq j} q_{jk}$ just to place it in the similar form as the discrete case. In general, we do not show self-loops on a state transition diagram as they are simply the negative sum of all rates leaving those states. On the contrary, self-loops on a state transition diagram for a discrete-time Markov chain indicate the probabilities of staying in those states and are usually shown on the diagram. Note that $\tilde{P}$ and $\tilde{Q}$ are related in the following expression:

$$\frac{d}{dt}\tilde{P}(t) = \tilde{P}(t)\tilde{Q} = \tilde{Q}(t)\tilde{P}(t)$$

It is not surprising to see the similarity in form between the above expression and that governing the state probabilities because it can be shown that the following limits always exist and are independent of the initial state of the chain for an irreducible homogeneous Markov chain:

$$\lim_{t \to \infty} p_{ij}(t) = \pi_j$$

and

$$\lim_{t \to \infty} \pi_j(t) = \pi_j$$

3.4 BIRTH-DEATH PROCESSES

A birth-death process is a special case of a Markov chain in which the process makes transitions only to the neighbouring states of its current position. This restriction simplifies the treatments and makes it an excellent model for all the basic queueing systems to be discussed later. In this section and subsequent chapters, we shall relax our terminology somewhat and use the term 'Markov process' to refer to a continuous-time Markov chain as called in most queueing literature. The context of the discussion should be clear whether we refer to a discrete state space or a continuous state space Markov process.

The term birth-death process originated from a modelling study on the changes in the size of a population. When the population size is k at time t, the process is said to be in state k at time t. A transition from the k to $K + 1$ state signifies a 'birth' and a transition down to $k - 1$ denotes a 'death' in the population. Implicitly implied in the model is the assumption that there are no simultaneous births or deaths in an incremental time period Δt. As usual, we represent the state space by a set of positive integer numbers $\{0, 1, 2, \ldots\}$ and the state transition-rate diagram is depicted in Figure 3.6.

If we define λ_k to be the birth rate when the population is of size k and μ_k to be the death rate when the population is k, then we have

$$\lambda_k = q_{k,k+1} \quad \mu_k = q_{k,k-1}$$

Substituting them into the transition-rate matrix (3.31), we have the transition-rate matrix as

$$Q = \begin{pmatrix} -\lambda_0 & \lambda_0 & 0 & 0 & 0 \\ \mu_1 & -(\lambda_1 + \mu_1) & \lambda_1 & 0 & 0 \\ 0 & \mu_2 & -(\lambda_2 + \mu_2) & \lambda_2 & 0 & \cdots \\ 0 & 0 & \mu_3 & -(\lambda_3 + \mu_3) & \lambda_3 & \cdots \\ \cdots & & & & \ddots & \cdots \end{pmatrix}$$

By expanding Equation (3.32) and using the more familiar notation $P_k(t) = \pi_k$, we have

$$\frac{d}{dt}P_k(t) = -(\lambda_k + \mu_k)P_k(t) + \lambda_{k-1}P_{k-1}(t) + \mu_{k+1}P_{k+1}(t) \quad k \geq 1 \qquad (3.34)$$

Figure 3.6 Transition diagram of a birth-death process

$$\frac{d}{dt} p_0(t) = -\lambda_0 p_0(t) + \mu_1 P_1(t) \quad k = 0 \tag{3.35}$$

In general, finding the time-dependent solutions of a birth-death process is difficult and tedious, and at times unmanageable. We will not pursue it further but rather show the solutions of some simple special cases. Under very general conditions and for most of the real-life systems, $P_k(t)$ approaches a limit P_k as $t \to \infty$ and we say the system is in statistical equilibrium.

Example 3.10

A special case of the birth-death process is the *pure-birth* process where $\lambda_k = \lambda > 0$ and $\mu_k = 0$ for all k. Assume that the initial condition is $P_0(0) = 0$, we have from Equations (3.31) and (3.32) the following equations:

$$\frac{dP_k(t)}{dt} = -\lambda P_k(t) + \lambda P_{k-1}(t) \quad k \geq 1$$

$$\frac{dP_0(t)}{dt} = -\lambda P_0(t)$$

It was shown in Chapter 1 that the solution that satisfies these set of equations is a Poisson distribution:

$$P_k(t) = \frac{(\lambda t)^k}{k!} e^{-\lambda t}$$

This gives us another interpretation of the Poisson process. It can now be viewed as a pure-birth process.

Example 3.11

A typical telephone conversation usually consists of a series of alternate talk spurts and silent spurts. If we assume that the length of these talk spurts and silent spurts are exponentially distributed with mean $1/\lambda$ and $1/\mu$, respectively, then the conversation can be modelled as a two-state Markov process.

Solution

Let us define two states; state 0 for talk spurts and state 1 for silent spurts. The state transition rate diagram is shown in Figure 3.7.

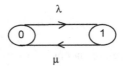

Figure 3.7 A two-state Markov process

The infinitesimal generator is given by $\tilde{Q} = \begin{pmatrix} -\lambda & \lambda \\ \mu & -\mu \end{pmatrix}$. Using expression (3.32) we have

$$\frac{d}{dt}P_0(t) = -\lambda P_0(t) + \mu P_1(t)$$

$$\frac{d}{dt}P_1(t) = \lambda P_0(t) - \mu P_1(t)$$

Let us define the Laplace transform of $P_0(t)$ and $P_1(t)$ as

$$F_0(s) = \int_0^\infty e^{-st} P_0(t)dt \quad \text{and} \quad F_1(s) = \int_0^\infty e^{-st} P_1(t)dt$$

and we have

$$sF_0(s) - P_0(0) = -\lambda F_0(s) + \mu F_1(s)$$

$$sF_1(s) - P_1(0) = \lambda F_0(s) - \mu F_1(s)$$

Let us further assume that the system begins in state 0 at $t = 0$, that is $P_0(0) = 1$ and $P_1(0) = 0$. Solving the two equations coupled with the initial condition, we have

$$F_0(s) = \frac{s+\mu}{s(s+\lambda+\mu)} = \frac{\mu}{\lambda+\mu} \cdot \frac{1}{s} + \frac{\lambda}{\lambda+\mu} \cdot \frac{1}{s+(\lambda+\mu)}$$

and

$$F_1(s) = \frac{\lambda}{s(s+\lambda+\mu)} = \frac{\lambda}{\lambda+\mu} \cdot \frac{1}{s} - \frac{\lambda}{\lambda+\mu} \cdot \frac{1}{s+(\lambda+\mu)}$$

Inverting the Laplace expressions, we obtain the time domain solutions as

$$P_1(t) = \frac{\lambda}{\lambda+\mu} - \frac{\lambda}{\lambda+\mu} e^{-(\lambda+\mu)t}$$

$$P_0(t) = \frac{\mu}{\lambda+\mu} + \frac{\lambda}{\lambda+\mu} e^{-(\lambda+\mu)t}$$

$$= \frac{\mu}{\lambda+\mu} + \left(1 - \frac{\mu}{\lambda+\mu}\right) e^{-(\lambda+\mu)t}$$

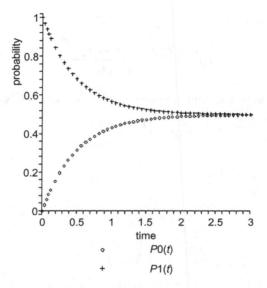

Figure 3.8   Probability distribution of a two-state Markov chain

For $\lambda = \mu = 1$, the plot of the two curves are shown in Figure 3.8. We see that both $p_0 = p_1 = 0.5$ when $t \to \infty$:

Example 3.12

A Yule process is a pure-birth process where $\lambda_k = k\lambda$ for $k = 0, 1, 2, \ldots$. Assuming that the process begins with only one member, that is $P_1(0) = 1$, find the time-dependent solution for this process.

Solution

Given the assumption of $\lambda_k = k\lambda$, we have the following equation from Equation (3.33):

$$\frac{dP_k(t)}{dt} = -\lambda k P_k(t) + \lambda(k-1)P_{k-1}(t) \quad k \geq 1$$

Define the Laplace transform of

$$P_k(t) \quad \text{as} \quad F_k(s) = \int_0^\infty e^{-st} P_k(t) dt$$

then we have

$$F_k(s) = \frac{P_k(0) + \lambda(k-1)F_{k-1}(s)}{s + k\lambda}$$

Since the system begins with only one member at time $t = 0$, we have

$$P_k(0) = \begin{cases} 1 & k = 1 \\ 0 & otherwise \end{cases}$$

Back substituting $F_k(s)$, we have

$$F_1(s) = \left(\frac{1}{s+\lambda}\right)$$

$$F_2(s) = \left(\frac{1}{s+\lambda}\right)\left(\frac{\lambda}{s+2\lambda}\right)$$

$$F_k(s) = \left(\frac{1}{s+\lambda}\right)\prod_{j=1}^{k-1}\left(\frac{j\lambda}{s+(j+1)\lambda}\right)$$

The expression can be solved by carrying out partial-fraction expansion of the right-hand side first and then inverting each term. Instead of pursuing the general solution of $P_k(t)$, let us find the time distribution of $P_2(t)$ and $P_3(t)$ just to have a feel of the probability distribution:

$$F_2(s) = \left(\frac{1}{s+\lambda}\right)\left(\frac{\lambda}{s+2\lambda}\right) = \frac{1}{s+\lambda} - \frac{1}{s+2\lambda}$$

$$F_3(s) = \left(\frac{1}{s+\lambda}\right)\left(\frac{\lambda}{s+2\lambda}\right)\left(\frac{2\lambda}{s+3\lambda}\right) = \frac{1}{s+\lambda} - \frac{2}{s+2\lambda} + \frac{1}{s+3\lambda}$$

Inverting these two expressions, we have the time domain solutions as

$$P_2(t) = e^{-\lambda t} - e^{-2\lambda t}$$
$$P_3(t) = e^{-\lambda t} - 2e^{-2\lambda t} + e^{-3\lambda t}$$

The plots of these two distributions are shown in Figure 3.9.

Problems

1. Consider a sequence of Bernoulli trials with probability of 'success' p, if we define X_n to be the number of uninterrupted successes that

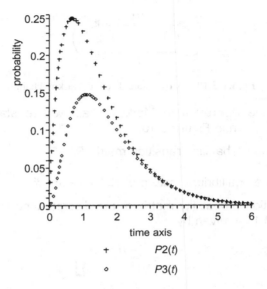

Figure 3.9 Probability distribution of a Yule process

have been completed at n trial; that is if the first 6 outcomes are 'S', 'F', 'S','S', 'S', 'F'; then $X_1 = 1$, $X_2 = 0$, $X_3 = 1$, $X_4 = 2$, $X_5 = 3$, $X_6 = 0$.

(i) argue that X_n is a Markov chain
(ii) draw the state transition diagram and write down the one-step transition matrix.

2. By considering a distributed system that consists of three processors (A, B and C), a job which has been processed in processor A has a probability p of being redirected to processor B and probability $(1 - p)$ to processor C for further processing. However, a job at processor B is always redirected back to process A after completion of its processing, whereas at processor C a job is redirected back to processor B q fraction of the time and $(1 - q)$ fraction of the time it stays in processor C.

(i) Draw the state transition diagram for this multi-processor system.
(ii) Find the probability that a job X is in processor B after 2 routings assuming that the job X was initially with processor A.
(iii) Find the steady-state probabilities.
(iv) Find value of p and q for which the steady state probabilities are all equal.

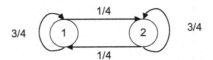

Figure 3.10   A two-state discrete Markov chain

3. Consider the discrete-time Markov chain whose state transition diagram is given in Figure 3.10.
   (i)   Find the probability transition matrix $\tilde{P}$.
   (ii) Find $\tilde{P}^5$.
   (iii) Find the equilibrium state probability vector $\tilde{\pi}$.

4. The transition probability matrix of the discrete-time counterpart of Example 3.11 is given by

$$\tilde{P} = \begin{bmatrix} 1-p & p \\ q & 1-q \end{bmatrix}$$

   Draw the Markov chain and find the steady-state vector for this Markov chain.

5. Consider a discrete-time Markov chain whose transition probability matrix is given by

$$\tilde{P} = \begin{bmatrix} 1 & 0 \\ 0 & 1 \end{bmatrix}$$

   (i)   Draw the state transition diagram.
   (ii) Find the k-step transition probability matrix $\tilde{P}^k$.
   (iii) Does the steady-state probability vector exist?

6. By considering the lift example 3.2, for what initial probability vector will the stationary probabilities of finding the life at each floor be proportional to their initial probabilities? What is this proportional constant?

7. A pure-death process is one in which members of the population only die but none are born. By assuming that the population begins with N members, find the transient solution of this pure death process.

8. Consider a population with external immigration. Each individual member in the population gives birth at an exponential rate λ and dies at an exponential rate μ. The external immigration is assumed to contribute to an exponential rate of increase θ of the population. Assume that the births are independent of the deaths as well as the external immigration. How do you model the population growth as a birth-death process?

4

Single-Queue Markovian Systems

In Chapter 3 we showed that it was easier to determine $N(t)$, the number of customers in the system, if it could be modelled as a Markov process. Here, we shall demonstrate the general approach employed in obtaining these performance measures of a Markovian queueing system using the Markov chain theory. We first model the number of customers present in a queueing system as a birth-death process and then compute the equilibrium state probability P_k using balance equations. Once P_k is found, we can then calculate N using the expression:

$$N = \sum_k kP_k$$

and then proceed to evaluate other parameters that are of interest by using Little's theorem.

A Markovian queueing system is one characterized by a Poisson arrival process and exponentially distributed service times. In this chapter we will examine only this group of queueing systems or those that can be adapted as such. These queueing models are useful in applications to data communications and networking. The emphasis here is on the techniques of computing these performance measures rather than to comprehensively cover all queueing systems. Students are referred to (Kleinrock 1975) which gives an elegant analysis for a collection of queueing systems.

We assume all the queueing systems we deal with throughout this book are ergodic and we shall focus our attention on the queueing results in the steady

Queueing Modelling Fundamentals Second Edition Ng Chee-Hock and Soong Boon-Hee
© 2008 John Wiley & Sons, Ltd

Figure 4.1   An M/M/1 system

state. The system is said to be in steady state when all transient behaviour has
subsided and the performance measures are independent of time.

4.1   CLASSICAL M/M/1 QUEUE

This classical M/M/1 queue refers to a queueing system where customers arrive
according to a Poisson process and are served by a single server with an exponential service-time distribution, as shown in Figure 4.1. The arrival rate λ and
service rate μ do not depend upon the number of customers in the system so
are state-independent. Recall that the defaults for the other two parameters in
Kendall's notation are infinite system capacity and first-come first-served
queueing discipline.

Firstly, let us focus our attention on the number of customers $N(t)$ in the
system at time t. A customer arriving at the system can be considered as a birth
and a customer leaving the system after receiving his service is deemed as a
death. Since the Poisson process prohibits the possibility of having more than
one arrival in Δt and the exponential service time ensures that there is at most
one departure in Δt, then clearly $N(t)$ is a birth-death process because it can
only go to its neighbouring states, $(N(t) + 1)$ or $(N(t) - 1)$ in a time interval
Δt.

From Section 3.4 we have the following expressions governing a birth-death
process and we can go on to find the time-dependent solution if so desired:

$$\frac{d}{dt}P_k(t) = -(\lambda_k + \mu_k)P_k(t) + \lambda_{k-1}P_{k-1}(t) + \mu_{k+1}P_{k+1}(t) \tag{4.1}$$

$$\frac{d}{dt}P_0(t) = -\lambda_0 P_0(t) + \mu_1 P_1(t) \tag{4.2}$$

Generally, in network performance evaluation we are more interested in the
long-term behaviour, that is the equilibrium state. If we assume a steady state
exists, then the rate of change of probability should go to zero, i.e.:

$$\frac{d}{dt}P_k(t) = 0 \quad \text{and} \quad P_k = \lim_{t \to \infty} P_k(t)$$

Since the birth and death rates are state independent, we shall drop the sub-
scripts, $\lambda_k = \lambda$ and $\mu_k = \mu$. And we have from Equations (4.1) and (4.2):

$$(\lambda + \mu)P_k = \lambda P_{k-1} + \mu P_{k+1} \tag{4.3}$$

$$\mu P_1 = \lambda P_0 \tag{4.4}$$

Equation (4.3) is a recursive expression and we shall solve it using z-transform. Multiplying the whole equation by z^k and summing from one to infinity, we have

$$\sum_{k=1}^{\infty}(\lambda + \mu)P_k z^k = \sum_{k=1}^{\infty}\lambda P_{k-1}z^k + \sum_{k=1}^{\infty}\mu P_{k+1}z^k$$

$$(\lambda + \mu)[P(z) - P_0] = \lambda z P(z) + \frac{\mu}{z}[P(z) - P_1 z - P_0]$$

Coupled with the boundary conditions of Equation (4.4), $P_1 = \dfrac{\lambda}{\mu}P_0$, we have

$$(\lambda + \mu)P(z) - \mu P_0 = \frac{\mu}{z}[P(z) - P_0] + \lambda z P(z)$$

$$P(z) = \frac{P_0 \mu(1-z)}{\lambda z^2 - (\lambda + \mu)z + \mu} = \frac{P_0 \mu(1-z)}{(\lambda z - \mu)(z - 1)}$$

$$= \frac{P_0}{1 - \dfrac{\lambda}{\mu}z}$$

$$= P_0 \frac{1}{1 - \rho z} \quad \text{where} \quad \rho = \lambda/\mu \tag{4.5}$$

To evaluate P_0, we use the normalization condition $\displaystyle\sum_{k=0}^{\infty} p_k = 1$, which is equivalent to $P(z)|_{z=1} = 1$. Setting $z = 1$ and $P(z) = 1$, we have

$$P_0 = 1 - \frac{\lambda}{\mu} = 1 - \rho \tag{4.6}$$

Hence we arrive at

$$P(z) = \frac{1 - \rho}{1 - \rho z} \tag{4.7}$$

and

$$P_k = (1 - \rho)\rho^k \tag{4.8}$$

We see that the probability of having k customers in the system is a geometric random variable with parameter ρ. It should be noted from the expression that ρ has to be less than unity in order for the sequence to converge and hence be a stable system.

4.1.1   Global and Local Balance Concepts

Before we proceed, let us deviate from our discussion of P_k and examine a very simple but powerful concept, *global balance* of probability flows. Recall the equilibrium stationary equation we derived in the preceding section for M/M/1:

$$(\lambda + \mu)P_k = \lambda P_{k-1} + \mu P_{k+1} \tag{4.9}$$

We know from probability theory that P_{k-1} is the fraction of time that the process found in state $(k - 1)$, therefore λP_{k-1} can be interpreted as the expected rate of transitions from state $(k - 1)$ to state k and this quantity is called the *stochastic* (or *probability*) *flow* from state $(k - 1)$ to state k.

Similarly, μP_{k+1} is the stochastic flow going from state $(k + 1)$ to state k. Thus, we see that the right-hand side of the equation represents the total stochastic flow into state k and the left-hand side represents the stochastic flow out of state k. In other words, Equation (4.9) tells us that the total stochastic flow in and out of a state should be equal under equilibrium condition, as shown in Figure 4.2. Equation (4.9) above is called the *global balance equation* for the Markov chain in question:

This concept of balancing stochastic flow provides us with an easy means of writing down these stationary equations by inspection without resorting to the expression (4.1), even for more general state-dependent queues discussed later in this chapter. In general, once we have the state transition diagram governing a continuous-time Markov chain, we can then write down the stationary

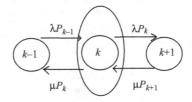

Figure 4.2   Global balance concept

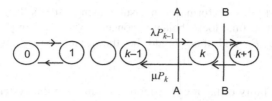

Figure 4.3 Local balance concept

equation by inspection using this flow balancing concept and so go on to solve the equation.

Extending our discussion of flow balancing further, let us consider an imaginary boundary B–B between node k and $k + 1$, as shown in Figure 4.3. If we equate the probability flows that go across this boundary, we have

$$\lambda P_k = \mu P_{k+1} \qquad (4.10)$$

This equation can be interpreted as a special case of the global balance in the sense that it equates the flow in and out of an imaginary super node that encompasses all nodes from node 0 to node k. This particular expression is referred as the *local balance equation* or *detailed balance equation*. It can be verified that this local balance equation always satisfies the global balance equation.

4.1.2 Performance Measures of the M/M/1 System

Coming back to our discussion of P_k and using the local balance equation across the boundary A–A of node $(k - 1)$ and node k, we have

$$P_k = \frac{\lambda}{\mu} P_{k-1} = \left(\frac{\lambda}{\mu}\right)^2 P_{k-2} = \left(\frac{\lambda}{\mu}\right)^3 P_{k-3}$$
$$= \rho^k P_0 \qquad (4.11)$$

As usual, we compute P_0 using the normalization equation $\sum_k P_k = 1$ to give

$$P_0 = (1 - \rho)$$

$$P_k = (1 - \rho)\rho^k$$

We obtain the same results as in Equations (4.6) and (4.8), without resorting to Markov theory and z-transform. Once we have obtained P_k, we can proceed to find other performance measures using Little's theorem. Students should note that this local balance concept offers a simpler expression for computing P_k and this should be the preferred approach as far as possible:

(i) The probability of having n or more customers in the system is given by

$$P[N \geq n] = \sum_{k=n}^{\infty} P_k$$

$$= (1-\rho) \sum_{k=n}^{\infty} \rho^k$$

$$= (1-\rho) \left[\sum_{k=0}^{\infty} \rho^k - \sum_{k=0}^{n-1} \rho^k \right]$$

$$= (1-\rho) \left[\frac{1}{1-\rho} - \frac{1-\rho^n}{1-\rho} \right]$$

$$= \rho^n \qquad\qquad (4.12)$$

(ii) The average number of customers N in the system in steady state is then given by

$$N = \sum_{k=0}^{\infty} kP_k = \sum_{k=0}^{\infty} k(1-\rho)\rho^k$$

$$= \rho(1-\rho) \sum_{k=0}^{\infty} k\rho^{k-1}$$

$$= \frac{\rho}{1-\rho} = \frac{\lambda}{\mu-\lambda} \qquad\qquad (4.13)$$

and the variance of N can be calculated as

$$\sigma_N^2 = \sum_{k=0}^{\infty} (k-N)^2 p_k = \frac{\rho}{(1-\rho)^2} \qquad\qquad (4.14)$$

Figure 4.4 shows the average number of customers as a function of utilization ρ. As the utilization (load) approaches the full capacity of the system ($\rho = 1$), the number of customers grows rapidly without limit and the system becomes unstable. Again, we see that for the system to be stable, the condition is $\rho < 1$.

This is a significant result for a single isolated queueing system and is the basis of the 'two-third' design rule. In practice, we usually design a shared

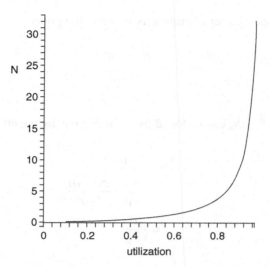

Figure 4.4 Number of customers in the system M/M/1

resource in such a way that its utilization is less than two-thirds of its full capacity.

(iii) The average time (T) a customer spends in the system is sometimes referred to as system time, system delay or queueing time in other queueing literature. We may use them interchangeably:

$$T = \frac{N}{\lambda} = \frac{\rho}{\lambda(1-\rho)}$$

$$= \frac{1}{\mu - \lambda} \tag{4.15}$$

(iv) The average customers at the service facility N_s:

$$N_s = \lambda/\mu = \rho$$

$$= 1 - P_0 \tag{4.16}$$

(iv) The average time a customer spends in the waiting queue is also known as *waiting time*. Students should not confuse this with the queueing time, which is the sum of waiting time and service time:

$$W = T - \frac{1}{\mu}$$

$$= \frac{\rho}{\mu - \lambda} \tag{4.17}$$

(vi) The average number of customers in the waiting queue:

$$N_q = \lambda W$$

$$= \frac{\rho^2}{1-\rho} \tag{4.18}$$

Alternatively, N_q can be found by the following argument:

$$N_q = N - \rho$$

$$= \frac{\rho}{1-\rho} - \frac{\rho(1-\rho)}{1-\rho}$$

$$= \frac{\rho^2}{1-\rho} \tag{4.19}$$

It should be noted that these steady-state performance measures are derived without any assumption about the queueing discipline and they hold for all queueing disciplines which are work conserving. Simply put, work conserving means that the server is always busy as long as there are customers in the system.

4.2 PASTA – POISSON ARRIVALS SEE TIME AVERAGES

Before we proceed, let us revisit the Poisson arrival process and examine a very useful and important concept called Poisson Arrivals See Time Averages (PASTA). This concept will be used below to obtain the distribution functions of system and waiting times.

Basically, this property states that for a stable queueing system with Poisson arrivals, the probability distribution as seen by an arriving customer is the same as the time-averaged (equilibrium or ergodic) state probability. In other words, the state $\{N(t) = k\}$ of the queueing system seen by an arrival has the same probability distribution as the state seen by an external random observer. That is

$$A_k(t) = P_k(t) \tag{4.20}$$

Here $A_k(t)$ is the probability of finding k customers in the system by an arriving customer and $P_k(t)$ is the steady-state probability of the system in state k. By definition:

$$A_k(t) \equiv P[N(t) = k | \text{An arrival at time } t]$$

$$= \frac{P[N(t) = k, \text{An arrival at time } t]}{P[\text{An arrival at time } t]}$$

$$= \frac{P[\text{An arrival at time } t | N(t) = k] \cdot P[N(t) = k]}{P[\text{An arrival at time } t]}$$

However, we know that the event of an arrival at time t is independent of the process $\{N(t) = k\}$ because of the memoryless property of Poisson process. That is

$$P[\text{An arrival at time } t \mid N(t) = k] = P[\text{An arrival at time } t]$$

Therefore, substituting in the preceding expression, we obtain

$$A_k(t) = P[N(t) = k] = P_k(t) \tag{4.21}$$

In the limit of $t \to \infty$, we have

$$A_k = P_k \tag{4.22}$$

4.3 M/M/1 SYSTEM TIME (DELAY) DISTRIBUTION

In previous sections, we obtained several performance measures of an M/M/1 queue, so we are now in a position to predict the long-range averages of this queue. However, we are still unable to say anything about the probability that a customer will spend up to 3 minutes in the system or answer any questions on that aspect. To address them, we need to examine the actual probability distribution of both the system time and waiting time. Let us define the system-time density function as

$$f_T(t) = \frac{d}{dt} P[T < t]$$

Students should note that the probability distributions (or density functions) of system time or waiting time depend very much on the actual service discipline used, although the steady state performance measures are independent of the service discipline.

We assume here that FCFS discipline is used and focus our attention on the arrival of customer i. The system time of this customer will be the sum of his/her service time and the service times of those customers (say k of them) ahead of him/her if the system is not empty when he/she arrives. Otherwise, his/her system time will be just his/her own service time. That is

$$T = \begin{cases} x_i + x_1 + x_2 + \ldots + x_k & k \geq 1 \\ x_i & k = 0 \end{cases} \tag{4.23}$$

Owing to the memoryless property of exponential service times, the remaining service time needed to finish serving the customer currently in service is still exponentially distributed. Hence the density function of T is simply the convolution of all the service times which are all exponentially distributed.

Using Laplace transform, the conditional system-time density function is

$$L[f_T(t|k)] = \left(\frac{\mu}{s+\mu}\right)^{k+1} \qquad k = 0, 1, 2, \ldots \tag{4.24}$$

Earlier we obtained $P_k = (1 - \rho)\rho^k$, hence the Laplace transform of the unconditional probability density function can be obtained using the total probability theorem:

$$\begin{aligned} L[f_T(t)] &= \sum_{k=0}^{\infty} \left(\frac{\mu}{s+\mu}\right)^{k+1} (1-\rho)\rho^k \\ &= \frac{\mu - \lambda}{s + (\mu - \lambda)} \end{aligned} \tag{4.25}$$

Inverting the above expression, we have

$$f_T(t) = (\mu - \lambda)e^{-(\mu-\lambda)t}, \quad t > 0 \tag{4.26}$$

and the cumulative distribution function $F_T(t)$ is

$$F_T(t) = P[T \leq t] = 1 - e^{-(\mu-\lambda)t} \tag{4.27}$$

At this juncture, it is important to point out that the above expression was actually derived with respect to an arriving customer. However, owing to the PASTA property of a Poisson arrival process, these probability distributions are the same as the long-range time averages or, in other words, distributions seen by a random observer.

Once the density function of system time is found, the density function of waiting time $f_w(t)$ can be found by considering:

$$T = w + x \tag{4.28}$$

Taking Laplace transform, we have

$$\frac{\mu - \lambda}{s + (\mu - \lambda)} = L[f_w(s)]\frac{\mu}{s + \mu}$$

$$L(f_w(s)) = \frac{(s + \mu)(1 - \rho)}{s + (\mu - \lambda)}$$

$$= (1 - \rho) + \frac{\lambda(1 - \rho)}{s + (\mu - \lambda)} \qquad (4.29)$$

Inverting

$$f_w(t) = (1 - \rho)\delta(t) + \lambda(1 - \rho)e^{-(\mu - \lambda)t} \qquad (4.30)$$

where $\delta(t)$ is the impulse function or Dirac delta function.

Integrating we have the waiting time cumulative distribution function $F_w(t)$:

$$F_w(t) = 1 - \rho e^{-(\mu - \lambda)t} \qquad (4.31)$$

Example 4.1

At a neighbourhood polyclinic, patients arrive according to a Poisson process with an average inter-arrival time of 18 minutes. These patients are given a queue number upon arrival and will be seen, by the only doctor manning the clinic, according to their queue number. The length of a typical consultation session is found from historical data to be exponentially distributed with a mean of 7 minutes:

(i) What is the probability that a patient has to wait to see the doctor?
(ii) What is the average number of patients waiting to see the doctor?
(iii) What is the probability that there are more than 5 patients in the clinic, including the one in consultation with the doctor?
(iv) What is the probability that a patient would have to wait more than 10 minutes for this consultation?
(v) The polyclinic will employ an additional doctor if the average waiting time of a patient is at least 7 minutes before seeing the doctor. By how much must the arrival rate increase in order to justify the additional doctor?

Solution

Assuming the clinic is sufficiently large to accommodate a large number of patients, the situation can be modelled as an M/M/1 queue. Given the following parameters:

$$\lambda = 1/18 \text{ person/min} \quad \text{and} \quad \mu = 1/7 \text{ person/min}$$

We have $\rho = 7/18$:

(i) $P[\text{patient has to wait}] = \rho = 1 - P_0 = 7/18$
(ii) The waiting-queue length $N_q = \rho^2 /(1 - \rho) = 49/198$
(iii) $P[N \geq 5] = \rho^5 = 0.0089$
(iv) Since $P[waitingtime \leq 10] = 1 - \rho e^{-\mu(1-\rho)t}$,

$$P[\text{queueing time} > 10] = \frac{7}{18} e^{-1/7(1-7/18)\times 10} = 0.162$$

(v)   Let the new arrival rate be λ'

then $\dfrac{7\lambda'}{\dfrac{1}{7} - \lambda'} \geq 7 \Rightarrow \lambda' = \dfrac{1}{14}$ person/min person/min.

Example 4.2

Variable-length data packets arrive at a communication node according to a Poisson process with an average rate of 10 packets per second. The single outgoing communications link is operating at a transmission rate of 64 kbits per second. As a first-cut approximation, the packet length can be assumed to be exponentially distributed with an average length of 480 characters in 8-bit ASCII format. Calculate the principal performance measures of this communication link assuming that it has a very large input buffer. What is the probability that 6 or more packets are waiting to be transmitted?

Solution

Again the situation can be modelled as an M/M/1 with the communication link being the server.

The average service time:

$$\bar{x} = \frac{480 \times 8}{64000} = 0.06\,\text{s}$$

and the arrival rate

$$\lambda = 10 \text{ packets/second}$$

thus

$$\rho = \lambda \bar{x} = 10 \times 0.06 = 0.6$$

The parameters of interest can be calculated easily as:

$$N = \frac{\rho}{1-\rho} = \frac{3}{2} \text{ packets} \quad \text{and} \quad N_q = \frac{\rho^2}{1-\rho} = \frac{9}{10} \text{ packets}$$

$$T = \frac{1}{\mu - \lambda} = \frac{1}{2} \text{ seconds} \quad \text{and} \quad W = \frac{\bar{x}}{\rho^{-1} - 1} = \frac{3}{10} \text{ seconds}$$

$P[\text{number of packets in the system} \geq 7]$

$= \rho^7 = (0.6)^7$

$= 0.028$

Example 4.3

In an ATM network, two types of packets, namely voice packets and data packets, arrive at a single channel transmission link. The voice packets are always accepted into the buffer for transmission but the data packets are only accepted when the total number of packets in the system is less than N.

Find the steady-state probability mass function of having k packets in the system if both packet streams are Poisson with rates λ_1 and λ_2, respectively. You may assume that all packets have exponentially distributed lengths and are transmitted at an average rate μ.

Solution

Using local balance concept (Figure 4.5), we obtain the following equations:

$$P_k = \begin{cases} \dfrac{\lambda_1 + \lambda_2}{\mu} P_{k-1} & k \leq N \\[2ex] \dfrac{\lambda_1}{\mu} P_{k-1} & k > N \end{cases}$$

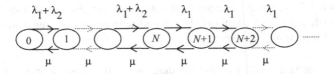

Figure 4.5 Transition diagram for Example 4.3

Using back substitution, we obtain

$$P_k = \left(\frac{\lambda_1 + \lambda_2}{\mu}\right)P_{k-1} = \left(\frac{\lambda_1 + \lambda_2}{\mu}\right)^2 P_{k-1} \quad \text{where} \quad \rho = \frac{\lambda_1 + \lambda_2}{\mu}$$

$$= \rho^k P_0$$

$$P_k = \left(\frac{\lambda_1}{\mu}\right)P_{k-1} = \left(\frac{\lambda_1}{\mu}\right)^2 P_{k-2} \quad \text{where} \quad \rho_1 = \frac{\lambda_1}{\mu}$$

$$= \rho_1^{k-N} P_N = \rho_1^{k-N} \rho^N P_0$$

To find P_0, we sum all the probabilities to 1:

$$P_0 \sum_{k=0}^{N} \rho^k + P_0 \sum_{k=N+1}^{\infty} \rho_1^{k-N} \rho^N = 1$$

$$P_0 \left\{ \sum_{k=0}^{N} \rho^k + \left(\frac{\rho}{\rho_1}\right)^N \left[\sum_{k=0}^{\infty} \rho_1^k - \sum_{k=0}^{N} \rho_1^k \right] \right\} = 1$$

$$P_0 = \left[\frac{1 - \rho^{N+1}}{1 - \rho} + \frac{\rho^N \rho_1}{1 - \rho_1} \right]^{-1}$$

Therefore, we have

$$P_k = \begin{cases} \rho^k \left[\dfrac{1 - \rho^{N+1}}{1 - \rho} + \dfrac{\rho^N \rho_1}{1 - \rho_1} \right]^{-1} \\[3mm] \rho_1^{k-N} \rho^N \left[\dfrac{1 - \rho^{N+1}}{1 - \rho} + \dfrac{\rho^N \rho_1}{1 - \rho_1} \right]^{-1} \end{cases}$$

Example 4.4

A computing facility has a small computer that is solely dedicated to batch-jobs processing. Job submissions get discouraged when the computer is heavily used and can be modelled as a Poisson process with an arrival rate $\lambda_k = \lambda/(k + 1)$ for $k = 0,1,2 \ldots$ when there are k jobs with the computer. The time taken to process each job is exponentially distributed with mean $1/\mu$, regardless of the number of jobs in the system.

(i)   Draw the state transitional-rates diagram of this system and write down the global as well as the local balance equation.

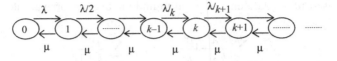

Figure 4.6 Transition diagram for Example 4.4

(ii) Find the steady-state probability P_k that there are k jobs with the computer and then find the average number of jobs.
(iii) Find the average time taken by a job from submission to completion.

Solution

(i) With reference to node k (Figure 4.6), we have

$$\text{Global:} \quad P_k\left(\mu + \frac{\lambda}{k+1}\right) = P_{k-1}\frac{\lambda}{k} + P_{k+1}\mu$$

With reference to the boundary between nodes $k - 1$ and k, we have

$$\text{Local:} \quad \frac{\lambda}{k}P_k = \mu P_{k-1}$$

(ii) From the local balance equation, we obtain

$$P_k = \frac{\rho^2}{k(k-1)}P_{k-2} = \frac{\rho^k}{k!}P_0 \quad \text{where} \quad \rho = \lambda/\mu$$

Summing all the probabilities to 1, we have

$$\sum_{k=0}^{\infty}\frac{\rho^k}{k!}P_0 = 1 \quad \Rightarrow \quad P_0 = e^{-\rho}$$

Therefore

$$P_k = \frac{\rho^k}{k!}e^{-\rho}$$

Since P_k is Poisson distributed, the average number of jobs $N = \rho$.

(iii) To use Little's theorem, we need to find the average arrival rate:

$$\bar{\lambda} = \sum_{k=0}^{\infty}\left(\frac{\lambda}{k+1}\right)\frac{\rho^k}{k!}e^{-\rho} = \mu e^{-\rho}\sum_{k=0}^{\infty}\frac{\rho^{k+1}}{(k+1)!}$$
$$= \mu e^{-\rho}(e^{\rho}-1) = \mu(1-e^{-\rho})$$

Therefore

$$T = \frac{N}{\bar{\lambda}} = \frac{\rho}{\mu(1-e^{-\rho})}$$

4.4   M/M/1/S QUEUEING SYSTEMS

The M/M/1 model discussed earlier is simple and useful if we just want to have a first-cut estimation of a system's performance. However, it becomes a bit unrealistic when it is applied to real-life problems, as most of them do have physical capacity constraints. Often we have a finite waiting queue instead of one that can accommodate an infinite number of customers. The M/M/1/S that we shall discuss is a more accurate model for this type of problem.

In *M/M/1/S*, the system can accommodate only *S* customers, including the one being served. Customers who arrive when the waiting queue is full are not allowed to enter and have to leave without being served. The state transition diagram is the same as the classical M/M/1 queue except that it is truncated at state *S*, as shown in Figure 4.7. This truncation of state transition diagram will affect the queueing results through P_0.

From last section, we have

$$P_k = \rho^k P_0 \qquad \text{where} \qquad \rho = \lambda/\mu$$

Using the normalization equation but sums to S state, we have

$$P_0\sum_{k=0}^{S}\rho^k = 1 \quad \Rightarrow \quad P_0 = \frac{1-\rho}{1-\rho^{S+1}} \tag{4.32}$$

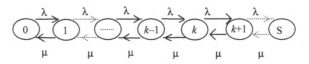

Figure 4.7   M/M/1/S transition diagram

$$P_k = \frac{(1-\rho)\rho^k}{1-\rho^{S+1}}$$ (4.33)

It should be noted that this system will always be ergodic and an equilibrium condition exists, even for the case where $\lambda \geq \mu$. This is due to the fact that the system has a self-regulating mechanism of turning away customers and hence the queue cannot grow to infinity. Students should note that the effective arrival rate that goes into the system is always less than μ.

4.4.1 Blocking Probability

Let us now digress from our discussion and look at the concept of *blocking probability P_b*. This is the probability that customers are blocked and not accepted by the queueing system because the system capacity is full. This situation occurs in queueing systems that have a finite or no waiting queue, hence it does not need to be an M/M/1/S queue. It can be any queueing system that blocks customers on arrival (Figure 4.8).

When the waiting queue of a system is full, arriving customers are blocked and turned away. Hence, the arrival process is effectively being split into two Poisson processes probabilistically through the blocking probability P_b. One stream of customers enters the system and the other is turned away, as shown in Figure 4.8.

The net arrival rate is

$$\lambda' = \lambda(1 - P_b).$$

For a stable system, the net departure rate γ should be equal to the net arrival rate, otherwise the customers in the system either will increase without bound or simply come from nowhere. Therefore

$$\gamma = \lambda(1 - P_b)$$

However, we know that the net departure rate can be evaluated by

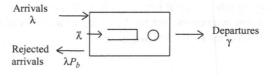

Figure 4.8 Blocking probability of a queueing

$$\gamma = \sum_{k=1}^{S} \mu P_k = \mu(1 - P_0) \tag{4.34}$$

Equating both expressions, we obtain

$$\lambda(1 - P_b) = \mu(1 - P_0)$$

$$P_b = \frac{P_0 + \rho - 1}{\rho} \tag{4.35}$$

The expression is derived for a queueing system with constant arrival and service rates; if they are state-dependent, then the corresponding ensemble averages should be used.

4.4.2 Performance Measures of M/M/1/S Systems

Continuing with our early discussion, the various performance measures can then be computed once P_k is found:

(i)   The saturation probability
By definition, the saturation probability is the probability when the system is full; that is there are S customers in the system. We say the system is saturated and we have

$$P_S = \frac{(1 - \rho)\rho^S}{1 - \rho^{S+1}}$$

However, substituting P_0 of M/M/1/S into expression (4.35), we have

$$P_b = \frac{P_0 + \rho - 1}{\rho} = \frac{(1 - \rho)\rho^S}{1 - \rho^{S+1}} \tag{4.36}$$

That is just the expression for P_s. This result is intuitively correct as it indicates that the blocking probability is the probability when the system is saturated. Owing to this blocking probability, the system now has a built-in self-regulatory mechanism and we see for the first time that it is still stable even when arrival rate is great than the service rate.

For $\rho = 1$, the expression for the blocking probability has to be evaluated using L'Hopital's rule and we have

$$P_b = \frac{1}{S + 1}$$

(ii) The average number of customers in the system is given by

$$N = \sum_{k=0}^{S} kP_k$$

$$= \sum_{k=1}^{S} \left(\frac{1-\rho}{1-\rho^{S+1}} \right) k\rho^k$$

$$= \left(\frac{1-\rho}{1-\rho^{S+1}} \right) \rho \sum_{k=1}^{s} k\rho^{k-1}$$

$$= \left(\frac{1-\rho}{1-\rho^{S+1}} \right) \rho \frac{d}{d\rho} \sum_{k=1}^{s} \rho^k$$

$$= \frac{\rho}{1-\rho} - \frac{(S+1)\rho^{S+1}}{1-\rho^{S+1}}$$

$$= \frac{\rho}{1-\rho} - \frac{\rho}{1-\rho}(S+1)P_S \tag{4.37}$$

Again, for $\rho = 1$, the expression has to be evaluated by L'Hopital's rule and we have

$$N = \frac{S}{2} \tag{4.38}$$

(iii) The average number of customers at the service facility:

$$N_s = P[k = 0]E[N_s|k = 0] + P[k > 0]E[N_s|N > 0]$$
$$= 1 - P_0$$
$$= \rho(1 - P_S) \tag{4.39}$$

(iv) The average number of customers in the waiting queue:

$$N_q = N - N_s = \frac{\rho^2}{1-\rho} - \frac{(S+\rho)\rho}{1-\rho}P_S \tag{4.40}$$

(v) The average time spent in the system and in the waiting queue.
Since customers are blocked when there are S customers in the system, the effective arrival rate of customers admitted into the system is

$$\lambda' = \lambda(1 - P_s) \tag{4.41}$$

and the T and W can be computed as

$$T = \frac{N}{\lambda'} = \frac{1}{\mu - \lambda} - \frac{S\rho^{S+1}}{\lambda - \mu\rho^{S+1}} \tag{4.42}$$

$$W = \frac{N_q}{\lambda'} = \frac{\rho}{\mu - \lambda} - \frac{S\rho^{S+1}}{\lambda - \mu\rho^{S+1}} \qquad (4.43)$$

Example 4.5

Variable-length packets arrive at a network switching node at an average rate of 125 packets per second. If the packet lengths are exponentially distributed with a mean of 88 bits and the single outgoing link is operating at 19.2 kb/s, what is the probability of buffer overflow if the buffer is only big enough to hold 11 packets? On average, how many packets are there in the buffer? How big should the buffer be in terms of number of packets to keep packet loss below 10^{-6}?

Solution

Given $\lambda = 125$ pkts/s $\qquad \mu^{-1} = 88/19200 = 4.6 \times 10^{-3}$ s
$\rho = \lambda\mu^{-1} = 0.573$

$$P_S = \frac{(1-\rho)\rho^S}{1-\rho^{S+1}} = \frac{(1-0.573)(0.573)^{12}}{1-(0.573)^{13}}$$

$$= 5.35 \times 10^{-4}$$

$$N_q = \frac{\rho^2}{1-\rho} - \frac{S+\rho}{1-\rho}P_d = 0.75$$

$$P_d = \frac{(1-0.573)(0.573)^S}{1-(0.573)^{S+1}} \le 10^{-6}$$

The above equation is best solved by trial and error. By trying various numbers, we have

$$S = 23 \ Pd = 1.2 \times 10^{-6}$$

$$S = 24 \ Pd = 6.7 \times 10^{-7}$$

Therefore, a total of 24 buffers is required to keep packet loss below 1 packet per million.

Example 4.6

Let us compare the performance of a statistical multiplexer with that of a frequency-division multiplexer. Assume that each multiplexer has 4 inputs of 5

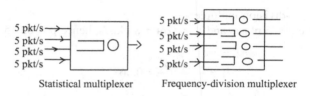

Figure 4.9 System model for a multiplexer

packets/second and a multiplexed output at 64 kbps. The statistical multiplexer has a buffer of 3 packets for the combined stream of input packets, whereas the frequency-division multiplexer has a buffer of 3 packets for each of the channels.

Solution

From the multiplexing principles, we see that the statistical multiplexer can be modelled as an M/M/1/4 queue and the equivalent frequency-division multiplexer as an 4 M/M/1/4 queue with the service rate of each server equal to a quarter of the statistical multiplexer, as shown in Figure 4.9.

(a) Statistical multiplexer

We have $\lambda = \sum_{i=1}^{4} \lambda_i = 20 \, \text{pkts/s}$ and $\mu = \dfrac{64000}{2000} = 32 \, \text{pkts/s}$

hence

$$\rho = 20/32 = 0.625$$

$$P_b = \frac{(1-\rho)\rho^4}{1-\rho^5} = 0.06325$$

$$N = \frac{\rho}{1-\rho} - \frac{\rho}{1-\rho}(S+1)P_b = 1.14 \text{ packets}$$

$$N_q = \frac{\rho^2}{1-\rho} - \frac{(S+\rho)\rho}{1-\rho}P_b = 0.555 \text{ packet}$$

Since there are 4 inputs, on average there are $(1.14/4) = 0.285$ packet per input in the system and $(0.555/4) = 0.1388$ packet per input in the buffer:

$$T = \frac{N}{\lambda(1-P_b)} = 0.06$$

$$W = \frac{N_q}{\lambda(1-P_b)} = 0.03$$

(b) Frequency-division multiplexer
   We have $\lambda = 5$ pkts/s and $\mu = 32/4 = 8$ pkts/s, hence $\rho = 0.625$

$$P_b = \frac{(1-\rho)\rho^4}{1-\rho^5} = 0.06325$$

$$N = \frac{\rho}{1-\rho} - \frac{\rho}{1-\rho}(S+1)P_b = 1.14$$

$$N_q = \frac{\rho^2}{1-\rho} - \frac{(s+\rho)\rho}{1-\rho}P_b = 0.555$$

$$T = \frac{N}{\lambda(1-P_b)} = 0.24$$

$$W = \frac{N_q}{\lambda(1-P_b)} = 0.118$$

We see that the number of packets per input in the system as well as in the waiting queue increases four-fold. The delays have also increased by four times.

4.5   MULTI-SERVER SYSTEMS – M/M/m

Having examined the classical single-server queueing system, it is natural for us now to look at its logical extension; the multi-server queueing system in which the service facility consists of m identical parallel servers, as shown in Figure 4.10. Here, identical parallel servers mean that they all perform the same functions and a customer at the head of the waiting queue can go to any of the servers for service.

If we again focus our attention on the system state $N(t)$, then $\{N(t),\ t \geq 0\}$ is a birth-death process with state-dependent service rates. When there is one customer, one server is engaged in providing service and the service rate is μ. If there are two customers, then two servers are engaged and the total service rate is 2μ, so the service rate increases until $m\mu$ and stays constant thereafter.

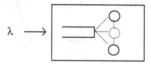

Figure 4.10   A multi-server system model

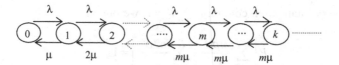

Figure 4.11   M/M/m transition diagram

There are several variations of multi-server systems. We shall first examine the *M/M/m* with an infinite waiting queue. Its corresponding state transition diagram is shown in Figure 4.11.

Using local balance concept, we can readily write down the governing equations by inspection:

$$k \le m \quad k\mu P_k = \lambda P_{k-1} \tag{4.44}$$

$$k \ge m \quad m\mu P_k = \lambda P_{k-1} \tag{4.45}$$

Recall in Chapter 2 that we defined the utilization as $\rho = \lambda/m\mu$ for a multi-server system. Hence, we have from Equation (4.44):

$$\begin{aligned}
P_k &= \frac{\lambda}{k\mu}P_{k-1} \\
&= \left(\frac{\lambda}{k\mu}\right)\left(\frac{\lambda}{(k-1)\mu}\right)P_{k-2} \\
&= \frac{(m\rho)^k}{k!}P_0
\end{aligned}$$

From Equation (4.45), we have

$$\begin{aligned}
P_k &= \frac{\lambda}{m\mu}P_{k-1}, \\
&= \left(\frac{\lambda}{\mu}\right)^{k-m}\left(\frac{1}{m}\right)^{k-m}P_m \\
&= \frac{(m\rho)^{k-m}}{m^{k-m}}\left(\frac{m\rho^m}{m!}P_0\right) \\
&= \frac{m^m\rho^k}{m!}P_0
\end{aligned}$$

Hence
$$P_k = \begin{cases} P_0\dfrac{(m\rho)^k}{k!} & k \le m \\[2ex] P_0\dfrac{m^m\rho^k}{m!} & k \ge m \end{cases} \tag{4.46}$$

Using the normalization condition $\sum P_k = 1$, we obtain

$$P_0 \left[\sum_{k=0}^{m-1} \frac{(m\rho)^k}{k!} + \sum_{k=m}^{\infty} \frac{m^m \rho^k}{m!} \right] = 1$$

$$P_0 = \left[\sum_{k=0}^{m-1} \frac{(m\rho)^k}{k!} + \frac{(m\rho)^m}{m!} \sum_{k=m}^{\infty} \rho^{k-m} \right]^{-1}$$

$$= \left[\sum_{k=0}^{m-1} \frac{(m\rho)^k}{k!} + \frac{(m\rho)^m}{m!(1-\rho)} \right]^{-1} \tag{4.47}$$

Similar to the *M/M/1* case, $\rho = \lambda/m\mu < 1$ is the condition for this system to be stable.

4.5.1   Performance Measures of M/M/m Systems

Once we have obtained P_k, as usual we can proceed to find other parameters. However, we usually express the performance measures of M/M/m systems in terms of a probability called *delay probability* or *queueing probability*, because this probability is widely used in designing telephony systems. It corresponds to the situation in classical telephony where no trunk is available for an arriving call. Its tabulated results are readily available and hence other parameters can be easily calculated.

(i)   The probability of delay
      This is the probability that an arriving customer finds all servers busy and is forced to wait in the queue. This situation occurs when there are more than m customers in the system, hence we have

$$P_d = P[delay] = \sum_{k=m}^{\infty} P_k$$

$$= \frac{P_o(m\rho)^m}{m!} \sum_{k=m}^{\infty} \rho^{k-m}$$

$$= \frac{P_0(m\rho)^m}{m!(1-\rho)} \tag{4.48}$$

This probability is often referred to as the Erlang C formula or the Erlang Delay formula and is often written as $C(\lambda/\mu, m)$. Most of the parameters of interest can be expressed in terms of this probability.

(ii) As the probability mass function P_k consists of two functions, it is easier to first find N_q, the number of customers waiting in the queue, instead of N so that the discontinuity in *pmf* can be avoided:

$$N_q = \sum_{k=0}^{\infty} kP_{m+k} = P_0 \frac{(m\rho)^m}{m!} \sum_{k=0}^{\infty} k\rho^k$$

$$= P_0 \frac{(m\rho)^m}{m!} \frac{\rho}{(1-\rho)^2}$$

$$= \frac{\rho}{1-\rho} P_d = \frac{\lambda}{m\mu - \lambda} P_d \tag{4.49}$$

(iii) The time spent in the waiting queue:

$$W = \frac{N_q}{\lambda} = \frac{1}{m\mu - \lambda} P_d \tag{4.50}$$

(iv) The time spends in the queueing system:

$$T = W + \frac{1}{\mu} = \frac{P_d}{m\mu - \lambda} + \frac{1}{\mu} \tag{4.51}$$

(v) The number of customers in the queueing system:

$$N = \lambda T = \frac{\rho}{1-\rho} P_d + m\rho \tag{4.52}$$

4.5.2   Waiting Time Distribution of M/M/m

Parallel to the M/M/1 case, we will continue with the examination of the waiting time distribution in this section. Again, we are going to make use of the PASTA concept to derive the function. Let us focus our attention on an arriving customer i to the system. He/her may find k customers waiting in the queue upon arrival, meaning there are $n = (k + m)$ customers in the system. He/she may also find no customers waiting in the queue, meaning the total number of customers is $n \leq m - 1$.

We denote the waiting time of this customers i as w, and its density function as $f_w(t)$ and cumulative probability function $F_w(t)$; then we have

$$F_w(t) = P[w < t]$$
$$= P[w < t, no\ waiting] + P[w < t, k \geq 0] \tag{4.53}$$

Let us examine each individual term on the right-hand side of the expression. The first term is simply the probability of having $n \leq m - 1$ customers in the system, hence no waiting:

$$P[w < t, no\ waiting] = P[n \le m - 1]$$
$$= \sum_{n=0}^{m-1} P_n = P_0 \sum_{n=0}^{m-1} \frac{(m\rho)^k}{k!} \qquad (4.54)$$

But, we have from Equation (4.47):

$$P_0 = \left[\sum_{k=0}^{m-1} \frac{(m\rho)^k}{k!} + \frac{(m\rho)^m}{m!(1-\rho)} \right]^{-1}$$

$$P_0 \sum_{k=0}^{m-1} \frac{(m\rho)^k}{k!} = 1 - \frac{P_0(m\rho)^m}{m!(1-\rho)}$$

Hence

$$P[w < t, no\ waiting] = 1 - \frac{P_0(m\rho)^m}{m!(1-\rho)} \qquad (4.55)$$

The second term depicts the situation where customer i has to wait in the queue. His/her waiting time is simply the sum of the service time x_j ($j = 1, \ldots k$) of those k customers in the queue as well as the service time x of those customers currently in service. That is

$$w = x + x_1 + x_2 + \ldots + x_k$$

The service times are all exponentially distributed with a rate $m\mu$, as there are m servers. Therefore the density function of w is simply the convolution of these service times, and in Laplace transform domain, the multiplication of them. Thus we have

$$L\{f_w(t|k)\} = \left(\frac{m\mu}{s+m\mu} \right)^{k+1}, \quad k = 0, 1, \ldots, \infty$$

and

$$L\{f_w(t)\} = \sum_{k=0}^{\infty} \left(\frac{m\mu}{s+m\mu} \right)^{k+1} \cdot P_{k+m} = \sum_{k=0}^{\infty} \left(\frac{m\mu}{s+m\mu} \right)^{k+1} \cdot P_0 \frac{m^m}{m!} \rho^{k+m}$$

$$= P_0 \frac{(m\rho)^m}{m!} \left(\frac{m\mu}{s+m\mu} \right) \sum_{k=0}^{\infty} \left(\frac{m\mu}{s+m\mu} \right)^k \cdot \rho^k$$

$$= P_0 \frac{(m\rho)^m}{m!} \left(\frac{m\mu}{s+m\mu} \right) \left(\frac{s+m\mu}{s+m\mu-\lambda} \right)$$

$$= P_0 \frac{(m\rho)^m}{m!} \left(\frac{m\mu}{s+m\mu-\lambda} \right) \qquad (4.56)$$

Inverting the Laplace transform, we have

$$f_w(t) = P_0 \frac{(m\rho)^m}{m!}(m\mu) \cdot e^{-(m\mu-\lambda)t} \qquad (4.57)$$

Integrating the density function, we obtain

$$F_w(t) = P_0 \frac{(m\rho)^m}{m!}\left(-\frac{m\mu}{m\mu - gl}\right) \cdot (e^{-(m\mu-\lambda)t} - 1)$$

$$= P_0 \frac{(m\rho)^m}{(1-\rho)m!} \cdot (1 - e^{-(m\mu-\lambda)t}) \qquad (4.58)$$

Combining both Equations (4.55) and (4.58), we have

$$F_w(t) = 1 - \frac{P_0(m\rho)^m}{m!(1-\rho)} + \frac{P_0(m\rho)^m}{(1-\rho)m!} \cdot (1 - e^{-(m\mu-\lambda)t})$$

$$= 1 - \frac{P_0(m\rho)^m}{(1-\rho)m!} e^{-(m\mu-\lambda)t}$$

$$= 1 - P_d e^{-(m\mu-\lambda)t} \qquad (4.59)$$

where P_d is the delay (queueing) probability. We see the similarity of this expression with that of the M/M/1 case, except now the ρ is replaced with P_d and μ with $m\mu$.

The time a customer spent in the system, system time or system delay, is simply his/her waiting time plus his/her own service time, i.e. $T = w + x$. Note that his/her service time is exponentially distributed with μ instead of $m\mu$ as the service rate of each individual server is μ. It can then be shown that the cumulative probability function of system time $F_T(t)$ is given by

$$F_T(t) = (1 - P_d)\mu e^{-\mu t} + \frac{P_d}{1 - m(1-\rho)}(1-\rho)m\mu(e^{-(m\mu-\lambda)t} - e^{-\mu t}) \qquad (4.60)$$

4.6 ERLANG'S LOSS QUEUEING SYSTEMS – M/M/m/m SYSTEMS

Similar to the single-server case, the assumption of infinite waiting queue may not be appropriate in some applications. Instead of looking at the multi-server system with finite waiting queue, we shall examine the extreme case M/M/m/m system where there is no waiting queue. This particular model has been widely used in evaluating the performance of circuit-switched telephone systems. It corresponds to the classical case where all the trunks of a telephone system are occupied and connection cannot be set up for further calls.

Since there is no waiting queue, when a customer arrives at the system and finds all servers engaged, the customer will not enter the system and is considered lost. The state transition diagram is the same as that of M/M/m except that it is truncated at state m.

Again using local balance concept, we have the following equation by inspection:

$$\lambda P_{k-1} = k\mu P_k$$

$$P_k = P_0 \frac{(m\rho)^k}{k!} \tag{4.61}$$

Solving for P_0:

$$P_0 = \left[\sum_{k=0}^{m} \frac{(m\rho)^k}{k!} \right]^{-1} \tag{4.62}$$

hence, we have

$$P_k = \frac{(m\rho)^k / k!}{\left[\sum_{k=0}^{m} \frac{(m\rho)^k}{k!} \right]} \tag{4.63}$$

Again this system is stable even for $\rho = \lambda/m\mu \geq 1$ because of this self-regulating mechanism of turning away customers when the system is full.

4.6.1   Performance Measures of the M/M/m/m

(i)   The probability that an arrival will be lost when all servers are busy:

$$P_b = \frac{(m\rho)^m / m!}{\sum_{k=0}^{m} (m\rho)^k / k!} = P_m \tag{4.64}$$

This probability is the blocking probability and is the same as the probability when the system is full. It is commonly referred to as the *Erlang B formula* or *Erlang's loss formula* and is often written as $B(\lambda/\mu, m)$.

(ii) The number of customers in the queueing system:

$$N = \sum_{k=0}^{m} kP_k = \sum_{k=1}^{m} k \cdot P_0 \frac{(m\rho)^k}{k!} = (m\rho) \sum_{k=0}^{m-1} P_0 \frac{(m\rho)^k}{k!}$$

$$= m\rho \left[\frac{\displaystyle\sum_{k=0}^{m} \frac{(m\rho)^k}{k!} - \frac{(m\rho)^m}{m!}}{\displaystyle\sum_{i=0}^{m} \frac{(m\rho)^i}{i!}} \right] = (m\rho)(1 - P_m) = m\rho(1 - P_b) \qquad (4.65)$$

(iii) The system time and other parameters

Since there is no waiting, the system time (or delay) is the same as service time and has the same distribution:

$$T = 1/\mu \qquad (4.66)$$

$$F_T(t) = P[T \le t] = 1 - e^{-\mu t} \qquad (4.67)$$

$$N_q = 0 \quad \text{and} \quad W = 0 \qquad (4\ 68)$$

4.7   ENGSET'S LOSS SYSTEMS

So far, we have always assumed that the size of the arriving customers or customer population is infinite, and hence the rate of arrivals is not affected by the number of customers already in the system. In this section, we will look at a case where the customer population is finite and see how it can be adapted to the birth-death process.

Engset's loss system is a 'finite population' version of Erlang's loss system. This is a good model for a time-sharing computer system with a group of fully-occupied video terminals. The jobs generated by each video terminal are assumed to be a Poisson process with rate λ. When a user at a video terminal has submitted a job, he is assumed to be waiting for the reply (answer) from the central CPU, and hence the total jobs' arrival rate is reduced.

In this mode, we again have m identical parallel servers and no waiting queue. However, we now have a finite population (C) of arriving customers instead of infinite population of customers, as shown in Figure 4.12.

To reflect the effect of the reduced arrival rate due to those customers already arrived at the system, the following parameters for the birth-death process are selected. The state transition diagram is shown in Figure 4.13.

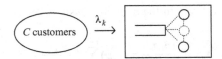

Figure 4.12   A M/M/m/m system with finite customer population

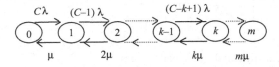

Figure 4.13   Transition diagram for M/M/m/m with finite customers

$$\lambda_k = \lambda(C - k) \quad 0 \le k \le C - 1$$
$$\mu_k = k\mu \qquad\qquad 0 \le k \le m \tag{4.69}$$

Again, using the local balance equation, we arrive at

$$(C - k + 1)\lambda P_{k-1} = k\mu P_k$$
$$\begin{aligned}
P_k &= \left(\frac{C - k + 1}{k}\right)\rho P_{k-1}, \qquad \rho = \lambda/\mu \\
&= \left(\frac{C - k + 1}{k}\right)\left(\frac{C - k + 2}{k - 1}\right)\rho^2 P_{k-2} \\
&= \left(\frac{C - k + 1}{k}\right)\left(\frac{C - k + 2}{k - 1}\right)\cdots\left(\frac{C}{1}\right)\rho^k P_0 \\
&= \binom{C}{k}\rho^k P_0 \tag{4.70}
\end{aligned}$$

Using the normalization condition to find P_0; we have

$$P_0 = \frac{1}{\displaystyle\sum_{k=0}^{m}\binom{C}{k}\rho^k} \tag{4.71}$$

Therefore:

$$P_k = \frac{\binom{C}{k}\rho^k}{\displaystyle\sum_{i=0}^{m}\binom{C}{i}\rho^i} \tag{4.72}$$

If the population size C of the arriving customers is the same as the number of servers in the system, i.e. $C = m$, then

$$P_k = \frac{\binom{m}{k}\rho^k}{\sum_{i=0}^{m}\binom{m}{i}\rho^i} = \frac{\binom{m}{k}\rho^k}{(1+\rho)^m} \tag{4.73}$$

4.7.1 Performance Measures of M/M/m/m with Finite Customer Population

(i) The probability of blocking; that is the probability of having all m servers engaged:

$$P_b = P_m = \frac{\binom{C}{m}\rho^m}{\sum_{k=0}^{m}\binom{C}{k}\rho^k} \tag{4.74}$$

(ii) The number of customers in the queueing system.

Instead of the usual approach, it is easier in this case to compute N by first calculating the average arrival rate $\bar{\lambda}$:

$$\bar{\lambda} = \sum_{k=0}^{m-1}\lambda_k P_k = \sum_{k=0}^{m-1}\lambda(C-k)\left[P_0\binom{C}{k}\rho^k\right]$$

$$= \lambda C\frac{\sum_{k=0}^{m-1}\binom{C}{k}\rho^k}{\sum_{k=0}^{m}\binom{C}{k}\rho^k} - \lambda\frac{\sum_{k=0}^{m-1}\binom{C}{k}\rho^k}{\sum_{k=0}^{m}\binom{C}{k}\rho^k}$$

$$= \lambda C\frac{\sum_{k=0}^{m}\binom{C}{k}\rho^k - \binom{C}{m}\rho^m}{\sum_{k=0}^{m}\binom{C}{k}\rho^k} - \lambda\frac{\sum_{k=0}^{m}k\binom{C}{k}\rho^k - m\binom{C}{m}\rho^m}{\sum_{k=0}^{m}\binom{C}{k}\rho^k}$$

$$= \lambda C(1-P_b) - \lambda(N - mP_b) \tag{4.75}$$

where N is the expected number of customers in the system and is given by the usual definition:

$$N = P_0\sum_{k=0}^{m}k\binom{C}{k}\rho^k \tag{4.76}$$

But we know that

$$N = \bar{\lambda}T = \bar{\lambda} \times \frac{1}{\mu}$$

therefore

$$N = \frac{1}{\mu}[\lambda C(1 - P_b) - \lambda(N - mP_b)]$$

$$N = \frac{\rho}{1+\rho}C - \frac{\rho}{1+\rho}(C - m)P_b \qquad (4.77)$$

where

$$\rho = \frac{\lambda}{\mu}$$

If the population size C of the arriving customers is the same as the number of servers in the system, i.e. $C = m$, then

$$N = \frac{\rho}{1+\rho}m \qquad (4.78)$$

(iii) The system time

$$T = \frac{1}{\mu} \qquad (4.79)$$

4.8   CONSIDERATIONS FOR APPLICATIONS OF QUEUEING MODELS

Those queueing models discussed so far are fairly elegant in analysis and yield simple closed-form results owing to the memoryless property of Poisson process and exponential service times. However, how valid are the assumptions of Poisson arrival process and exponential service times in real-life applications?

From the past observation and studies, the Poisson arrival has been found to be fairly accurate in modelling the arrival of calls in a telephone network, and hence was extensively used in designing such networks. It is a fairly good model as long as the arriving customer population is large and there is no interdependency among them on arrival. It may not fully match the arrival process of some other real-life problems, but so long as the discrepancy is small, it can be treated as the first-cut approximation.

The exponential service time assumption appears to be less ideal than the Poisson assumption but still offers fairly good results as far as voice networks are concerned. However, it may be inadequate in modelling packets or messages in data networks as the length of a packet is usually constrained by the physical implementation. The length could even be a constant, as in the case of ATM (Asynchronous Transfer Mode) networks. Nevertheless, the exponential distribution can be deemed as the worst-case scenario and will give us the first-cut estimation.

The M/M/1 queue and its variants can be used to study the performance measure of a switch with input buffer in a network. In fact, its multi-server counterparts have traditionally been employed in capacity planning of telephone networks. Since A K Erlang developed them in 1917, the M/M/m and M/M/m/m models have been extensively used to analyse the 'lost-calls-cleared' (or simply blocked-calls) and 'lost-calls-delayed' (queued-calls) telephone systems, respectively. A 'queued-calls' telephone system is one which puts any arriving call requests on hold when all the telephone trunks are engaged, whereas the 'blocked-calls' system rejects those arriving calls.

In a 'blocked-calls' voice network, the main performance criterion is to determine the probability of blocking given an offered load or the number of trunks (circuits) needed to provide certain level of blocking. The performance of a 'queued-calls' voice network is characterized by the Erlang C formula and its associated expressions.

Example 4.6

A trading company is installing a new 300-line PBX to replace its old existing over-crowded one. The new PBX will have a group of two-way external circuits and the outgoing and incoming calls will be split equally among them.

It has been observed from past experience that each internal telephone usually generated (call or receive) 20 minutes of voice traffic during a typical busy day. How many external circuits are required to ensure a blocking probability of 0.02?

Solution

In order for the new PBX to handle the peak load during a typical busy hour, we assume that the busy hour traffic level constitutes about 14% of a busy day's traffic.

Hence the total traffic presented to the PBX:

$$= 300 \times 20 \times 14\% \div 60$$
$$= 14 erlangs$$

The calculated traffic load does not account for the fact that trunks are tied up during call setups and uncompleted calls. Let us assume that these amount to an overhead factor of 10%.

Then the adjusted traffic = $14 \times (1 + 10\%) = 15.4 erlangs$

Using the Erlang B formula:

$$P_m = \frac{(15.4)^m / m!}{\sum_{k=0}^{m} (15.4)^k / k!} \leq 0.02$$

Again, we solve it by trying various numbers for m and we have

$$m = 22 \quad P_m = 0.0254$$
$$m = 23 \quad P_m = 0.0164$$

Therefore, a total of 23 lines is needed to have a blocking probability of less than or equal to 0.02.

Example 4.7

At the telephone hot line of a travel agency, information enquiries arrive according to a Poisson process and are served by 3 tour coordinators. These tour coordinators take an average of about 6 minutes to answer an enquiry from each potential customer.

From past experience, 9 calls are likely to be received in 1 hour in a typical day. The duration of these enquiries is approximately exponential. How long will a customer be expected to wait before talking to a tour coordinator, assuming that customers will hold on to their calls when all coordinator are busy? On the average, how many customers have to wait for these coordinators?

Solution

The situation can be modelled as an M/M/3 queue with

$$\rho = \frac{\lambda}{m\mu} = \frac{9/60}{3 \times (1/6)} = 0.3$$

$$P = \left[\sum_{k=0}^{m-1} \frac{(m\rho)^k}{k!} + \frac{(m\rho)^m}{m!} \left(\frac{1}{1-\rho} \right) \right]^{-1}$$

$$= \left[\sum_{k=0}^{2} \frac{(0.9)^k}{k!} + \frac{(0.9)^3}{3!} \left(\frac{1}{1-0.3} \right) \right]^{-1}$$

$$= 0.4035$$

$$P_d = P_0 \frac{(m\rho)^m}{m!} \left(\frac{1}{1-\rho} \right)$$

$$= 0.4035 \times \frac{(0.9)^3}{3!} \times \frac{1}{1-0.3}$$

$$= 0.07$$

$$N_q = \frac{\rho}{1-\rho} P_d = 0.03$$

$$W = \frac{N_q}{\lambda} = \frac{0.03}{9/60} = 0.2 \text{ minute}$$

Example 4.8

A multi-national petroleum company leases a certain satellite bandwidth to implement its mobile phone network with the company. Under this implementation, the available satellite bandwidth is divided into N_v voice channels operating at 1200 bps and N_d data channels operating at 2400 bps. It was forecast that the mobile stations would collectively generate a Poisson voice stream with mean 200 voice calls per second, and a Poisson data stream with mean 40 messages per second. These voice calls and data messages are approximately exponentially distributed with mean lengths of 54 bits and 240 bits, respectively. Voice calls are transmitted instantaneously when generated and blocked when channels are not available, whereas data messages are held in a large buffer when channels are not available:

(i) Find N_v, such that the blocking probability is less than 0.02.
(ii) Find N_d, such that the mean message delay is less than 0.115 second.

Solution

(i) $\lambda_v = 200$ and $\mu_v^{-1} = 54/1200 = 9/200$
     $(\lambda_v/\mu_v) = 9$

$$P_m = \frac{(\lambda_v/\mu_v)^m/m!}{\sum\limits_{k=0}^{m}(\lambda_v/\mu_v)^k/k!}$$

$$\frac{(9)^{N_v}/N_v!}{\sum\limits_{k=0}^{N_v}(9)^k/k!} \leq 0.02 \text{ therefore } N_v \geq 15$$

(ii) $\lambda_d = 40$ and $\mu_d^{-1} = 240/2400 = 0.1$
$(\lambda_d/\mu_d) = 4$

$$T = \frac{1}{\mu_d} + \frac{P_d}{m\mu_d - \lambda_d}$$

$$\frac{1}{10} + \frac{P_d}{10N_d - \lambda_d} \leq 0.115 \text{ and } N_d \geq 6$$

Example 4.9

A group of 10 video display units (VDUs) for transactions processing gain access to 3 computers ports via a data switch (port selector), as shown in Figure 4.14. The data switch merely performs connections between those VDUs and the computer ports.

If the transaction generated by each VDU can be deemed as a Poisson stream with rates of 6 transactions per hour, the length of each transaction is approximately exponentially distributed with a mean of 5 minutes. Calculate:

(i) The probability that all three computer ports are engaged when a VDU initiates a connection;
(ii) The average number of computer ports engaged;

It is assumed that a transaction initiated on a VDU is lost and will try again only after an exponential time of 10 minutes if it can secure a connection initially.

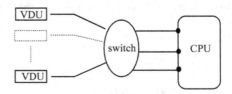

Figure 4.14   A VDU-computer set up

Solution

In this example, computer ports are servers and transactions are customers and the problem can be formulated as an M/M/m/m system with finite arriving customers.

Given $\lambda = 6/60 = 0.1$ trans/min $\quad \mu^{-1} = 5$ min $\quad$ therefore $\quad a = 0.5$

(i)
$$P_b = \frac{\binom{10}{3}(0.5)^3}{\displaystyle\sum_{k=0}^{3}\binom{10}{k}(0.5)^k} = 0.4651$$

(ii)
$$N = \frac{0.5}{1+0.5}\times 10 - \frac{0.5}{1+0.5}(10-3)\times 0.4651$$
$$= 2.248$$

Problems

1. By referring to Section 4.1, show that the variance of N and W for an M/M/1 queue are

   (i) $Var[N] = \rho/(1-\rho)^2$
   (ii) $Var[W] = [\mu(1-\rho)]^{-2}$

2. Show that the average number of customers in the M/M/1/S model is $N/2$ when $\lambda = \mu$.

3. Consider an M/M/S/S queueing system where customers arrive from a fixed population base of S. This system can be modelled as a Markov chain with the following parameters:

   $$\lambda_k = (s-k)\lambda$$
   $$\mu_k = k\mu$$

   Draw the state-transition diagram and show that the probability of an empty system is $(1 + \rho)^{-S}$. Hence find the expected number of customers in the system.

4. Data packets arrive at a switching node, which has only a single output channel, according to a Poisson process with rate λ. To handle congestion issues, the switching node implements a simple strategy of dropping incoming packets with a probability p when the total number of packets in the switch (including the one under transmission) is greater or more than N. Assuming that the

transmission rate of that output channel is μ, find the probability that an arriving packet is dropped by the switch.

5. A trading company intends to install a small PBX to handle ever-increasing internal as well as external calls within the company. It is expected that the employees will collectively generate Poisson out-going external calls with a rate of 30 calls per minute. The duration of these outgoing calls is independent and exponentially distributed with a mean of 3 minutes. Assuming that the PBX has separate external lines to handle the incoming external calls, how many exter-nal outgoing lines are required to ensure that the blocking probability is less than 0.01? You may assume that when an employee receives the busy tone he/she will not make an attempt again.

6. Under what conditions is the assumption 'Poisson arrival process and exponential service times' a suitable model for the traffic offered to a communications link? Show that when 'Poisson arrival process and exponential service times' traffic is offered to a multi-channel communications link with a very large number of channels, the equi-librium carried traffic distribution (state probability distribution) is Poisson. What is the condition for this system to be stable?

7. Consider Example 4.1 again. If a second doctor is employed to serve the patients, find the average number of patients in the clinic, assum-ing that the arrival and service rates remained the same as before?

8. A trading firm has a PABX with two long-distance outgoing trunks. Long-distance calls generated by the employees can be approxi-mated as a Poisson process with rate λ. If a call arrives when both trunks are engaged, it will be placed on 'hold' until one of the trunks is available. Assume that long-distance calls are exponentially dis-tributed with rate μ and the PABX has a large enough capacity to place many calls on hold:

(i) Show that the steady-state probability distribution is $\frac{2(1-\rho)}{1+\rho}\rho^k$

(ii) Find the average number of calls in the system.

9. By considering the Engset's loss system, if there is only one server instead of m servers, derive the probability of having k customers in the system and hence the average number of customers, assuming equilibrium exists.

5

Semi-Markovian
Queueing Systems

The queueing models that we have discussed so far are all Markovian types; that is both the arrival and service processes are memoryless. These models are appropriate in certain applications but may be inadequate in other instances. For example, in voice communication, the holding time of a call is approximately exponentially distributed, but in the case of packet switching, packets are transmitted either with a fixed length or a certain limit imposed by the physical constraints. In the latter case, a more appropriate model would be the M/G/1, which belongs to the class of queueing systems called *semi-Markovian queueing systems*.

Semi-Markovian queueing systems refer to those queueing systems in which either the arrival or service process is not 'memoryless'. They include M/G/1, G/M/1, their priority queueing variants and, of course, the multi-server counterparts.

In a Markovian queueing system, the stochastic process $\{N(t), t \geq 0\}$, that is the number of customers in the system at time t, summarizes the complete past history of the system and can be modelled as a birth-death process. This enables us to write down the balance equations from its state transition diagram, and then proceed to calculate P_k and its performance measures.

However, in the case of a semi-Markovian system, say M/G/1, we have to specify the service time already received by the customer in service at time t in addition to the $N(t)$. This is necessary as the distribution of the remaining service time for that customer under service is no longer the same as the original distribution. The process $N(t)$ no longer possesses the 'memoryless' property that was discussed in Chapter 2. The concept of flow balancing breaks down and we cannot set up simple balance equations for the process $\{N(t), t \geq 0\}$ using a transition rate diagram.

There are several techniques for solving this type of queueing system. The most frequently presented method in queueing literature is the *imbedded Markov-chain* approach in which we look at the queue behavior at those instants of a service completion, that is when a customer has finished receiving his/her service and left the system. In so doing, we get rid of the remaining service time to the completion of a service interval and again the system can be modelled by a birth-death process. Another approach is the so-called Residual Service Time method which examines the system process from an arriving customer's perspective. This method is simpler but can only give the mean value of the performance measures.

In this chapter, we present both approaches to highlight the various concepts and theories used in obtaining the results.

5.1   THE M/G/1 QUEUEING SYSTEM

This is a queueing system, Figure 5.1, where customers arrive according to a Poisson process with mean λ and are served by a single server of general service-time distribution $X(t)$ (and density function $x(t)$). We assume that the mean $E(x)$ and the second moment $E(x^2)$ of the service time distribution exist and are finite. We denote them as $\bar{x}$ and $\overline{x^2}$, respectively.

The capacity of the waiting queue is as usual infinite and customers are served in the order they arrived, i.e. FCFS. Note that this service discipline will not affect the mean queueing results providing it is work conserving. Simply put, work conserving means that the server does not stay idle when there are customers waiting in the queue. We also assume that the service times are independent of the arrival process as well as the number of customers in the system.

5.1.1   The Imbedded Markov-Chain Approach

As mentioned in the introduction, the stochastic process $N(t)$ is no longer sufficient to completely summarize the past history, as additional information such

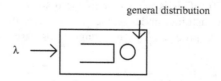

Figure 5.1   A M/G/1 queueing system

as the elapsed service time $x_0(t)$ needs to be specified. If we also specify the elapsed service time, then the tuple $[N(t), x_0(t)]$ is again a Markov process and we will be able to proceed with the analysis using Markov theory. Unfortunately, this is a two-dimensional specification of the state space and complicates the analysis considerably.

However, we can simplify the $[N(t), x_0(t)]$ specification into a one-dimensional description $N(t)$ by selecting certain special observation instants in time. If we observe the M/G/1 queue at departure instants $\{r_n\}$, where $x_0(t) = 0$, then we have a one-dimensional state space specification. The evolution of the number of customers $N(t)$ left behind by a departing customer at these instants is an imbedded Markov chain and we can again resort to Markov theory to derive the desired performance measures.

But before we proceed, let us take a closer look at the state description $N(t)$ at those departure epochs. This is the state probability that is seen by a departing customer. Is it the same as the limiting steady-state probability which is the system state observed by a random observer?

Fortunately, it can be shown that the system state seen by a departing customer is the same as the system state seen by an arriving customer. Kleinrock (1975) reasoned that the system state in an M/G/1 queue can change at most by +1 or −1. The former corresponds to a customer arrival and the latter refers to a departure. In the long term, the number of transitions upwards must equal the number of transitions downwards. Hence the system state distribution seen by a departing customer should be the same as that seen by an arriving customer.

And according to the PASTA property, as shown in Section 4.2, the system state distribution seen by an arriving customer is the same as the limiting steady-state distribution. We can therefore conclude that the system state distribution that we are going to derive with respect to a departing customer is the same as the limiting steady-state distribution seen by a random observer, that is the usual limiting state distribution we have been deriving for other queues.

With the assurance that the imbedded Markov process will yield the same limiting steady-state distribution, we can then proceed with our analysis.

5.1.2   Analysis of M/G/1 Queue Using Imbedded Markov-Chain Approach

Let us focus our attention at the departure epochs and examine the number of customers N_n left behind by customer C_n at the departure instant r_n. If C_n left behind a non-empty system, then customer C_{n+1} will leave behind a system with the number of customers increased by the number of arrivals during the service time of customer C_{n+1} minus themselves; that is $N_{n+1} = N_n - 1 + A_{n+1}$, where A_{n+1} is the number of arrivals during the service time of customer C_{n+1}.

However, if C_n leaves behind an empty queue, then the service does not start until C_{n+1} arrives. The number of customers left behind will merely be the number of arrivals during his/her service time, that is $N_{n+1} = A_{n+1}$. Combining both scenarios, we have

$$N_{n+1} = (N_n - 1)^+ + A_{n+1} \qquad (5.1)$$

where we have used the notation $(x)^+ = \max(x, 0)$. Note that A_{n+1} only depends upon the length of the service time (x_{n+1}) of C_{n+1} and not upon n at all, hence we can drop the subscript n. If we denote the probability of k arrivals in the service period of a typical customer as

$$a_k = P[A = k] \qquad (5.2)$$

then on condition that $x = t$ and using the law of total probability together with the fact that A is Poisson distributed with parameter λt as the arrival process of M/G/1 queue is Poisson, we have

$$a_k = P[A = k] = \int_0^\infty P[A = k | x = t] x(t) dt$$
$$= \int_0^\infty \frac{(\lambda t)^k}{k!} e^{-\lambda t} x(t) d(t) \qquad (5.3)$$

By definition, the transition probabilities of this imbedded Markov chain are given as

$$P_{ij} \triangleq P[N_{n+1} = j | N_n = i] \qquad (5.4)$$

Since these transitions are observed at departure instants, $N_{n+1} < N_n - 1$ is an impossible event and $N_{n+1} \geq N_n - 1$ is possible for all values due to the arrivals A_{n+1}, we have

$$P_{ij} = \begin{cases} a_{j-i+1} & i > 0, j \geq i-1 \\ a_j & i = 0, j \geq 0 \end{cases} \qquad (5.5)$$

And the transition probability matrix $\tilde{P} = [P_{ij}]$ is given as

$$\tilde{P} = \begin{bmatrix} a_0 & a_1 & a_2 & a_3 & \cdots \\ a_0 & a_1 & a_2 & a_3 & \cdots \\ 0 & a_0 & a_1 & a_2 & \cdots \\ 0 & a_0 & a_1 & a_2 & \cdots \\ \vdots & \vdots & \vdots & \vdots & \ddots \end{bmatrix} \qquad (5.6)$$

Given the service distribution, the transition probabilities are completely defined by Equation (5.3) and theoretically the steady state distribution can be found using $\tilde{\pi} = \tilde{\pi}\tilde{P}$. Instead of pursuing this line of analysis, we derive the system state using generating functions. We define the generating function of A as

$$A(z) \triangleq E[z^A] = \sum_{k=0}^{\infty} P[A=k]z^k \tag{5.7}$$

Using Equation (5.3), we have

$$\begin{aligned}
A(z) &= \sum_{k=0}^{\infty} \left\{ \int_0^{\infty} \frac{(\lambda t)^k}{k!} e^{-\lambda t} x(t)dt \right\} z^k \\
&= \int_0^{\infty} e^{-\lambda t} \left(\sum_{k=0}^{\infty} \frac{(\lambda t z)^k}{k!} \right) x(t)dt \\
&= \int_0^{\infty} e^{-(\lambda - \lambda z)t} x(t)dt \\
&= x^*(\lambda - \lambda z)
\end{aligned} \tag{5.8}$$

where $x^*(\lambda - \lambda z)$ is the Laplace transform of the service time pdf, $x(t)$, evaluated at $\lambda - \lambda z$.

Expression (5.8) reveals an interesting relationship between the number of arrivals occurring during a service interval x where the arrival process is Poisson at a rate of λ.

Let us evaluate the mean and second moment of A as we need them for the subsequent analysis of system state. Using Equation (5.8), we have

$$\begin{aligned}
\bar{A} \triangleq E[A] &= \frac{dA(z)}{dz} \bigg|_{z=1} \\
&= \frac{dx^*(\lambda - \lambda z)}{dz} \bigg|_{z=1} \\
&= \frac{dx^*(\lambda - \lambda z)}{d(\lambda - \lambda z)} \cdot \frac{d(\lambda - \lambda z)}{dz} \bigg|_{z=1} \\
&= -\lambda \cdot \frac{dx^*(\lambda - \lambda z)}{d(\lambda - \lambda z)} \bigg|_{z=1}
\end{aligned} \tag{5.9}$$

But the last term is just the mean service time $\bar{x}$, hence we arrive at

$$\bar{A} = \lambda \bar{x} = \rho \tag{5.10}$$

Proceeding with the differentiation further, we can show that

$$\overline{A}^2 \triangleq E[A^2]$$
$$= \lambda^2 E[x^2] + \overline{A} \tag{5.11}$$

5.1.3 Distribution of System State

Let us now return to our earlier investigation of the system state. We define the z-transform of the random variable N_n as

$$N_n(z) \triangleq E[z^{N_n}] = \sum_{k=0}^{\infty} P[N_n = k]z^k \tag{5.12}$$

Taking z-transform of Equation (5.1), we have

$$E[z^{N_{n+1}}] = E[z^{(N_n-1)^+ + A}] \tag{5.13}$$

We drop the subscript of A as it does not depend on n. Note that A is also independent of the random variable N_n, so we can rewrite the previous expression as

$$E[z^{N_{n+1}}] = E[z^{(N_n-1)^+}]E[z^A]$$

and combining the definition of their z-transform, we have

$$N_{n+1}(z) = A(z) \cdot E[Z^{(N_n-1)^+}] \tag{5.14}$$

Let us examine the second term on the right-hand side of the expression:

$$E[z^{(N_n-1)^+}] = \sum_{k=0}^{\infty} P[N_n = K]z^{(k-1)^+}$$

$$= P[N_n = 0] + \sum_{k=1}^{\infty} P[N_n = K]z^{k-1}$$

$$= (1-\rho) + \frac{1}{z}\sum_{k=0}^{\infty} P[N_n = K]z^k - \frac{1}{z}P[N_n = 0]$$

$$= (1-\rho) + \frac{1}{z}N_n(z) - \frac{1-\rho}{z} \tag{5.15}$$

We assume that the limiting steady-state exists, that is

$$N(z) = \lim_{n\to\infty} N_n(z) = \lim_{n\to\infty} N_{n+1}(z)$$

We then arrive at

$$N(z) = (1-\rho)\frac{(z-1)A(z)}{z-A(z)}$$

$$= (1-\rho)\frac{(z-1)x^*(\lambda-\lambda z)}{z-x^*(\lambda-\lambda z)} \tag{5.16}$$

Having obtained the generating function of the system state given in Equation (5.16), we can proceed to find the mean value N as

$$N = E[N(t)] = \frac{d}{dz}N(z)|_{z=1}$$

With N we can proceed to calculate other performance measures of the M/G/1. We will postpone this until we examine the Residual Service Time approach.

5.1.4   Distribution of System Time

System Time refers to the total time a typical customer spends in the system. It includes the waiting time of that customer and its service time. Note that though the distribution of system state derived in the previous sections does not assume any specific service discipline, the system time distribution is dependent on the order in which customers are served. Here, we assume that the service discipline is FCFS. The Laplace transform of the system time distribution $T(s)$ can be found using the extension of the concept adopted in the previous section. We will briefly describe it as this is a commonly used approach in analysing the delays of queueing systems.

Recall that in Section 5.1.2 we derived the generating function of the total arrivals A during a service time interval x, where the arrival process is Poisson with rate λ given by $A(z) = x^*(\lambda - \lambda z)$. Similarly, since a typical customer spends a system time T in the system, then the total arrivals during their system time interval will be given by

$$A_T(z) = \sum_{k=0}^{\infty}\left\{\int_0^\infty \frac{(\lambda t)^k}{k!}e^{-\lambda t}f_T(t)dt\right\}z^k$$

$$= T^*(\lambda - \lambda z) \tag{5.17}$$

where $A_T(z)$ is the generating function of the total arrivals, $f_T(t)$ is the density function of system time and $T^*(s)$ is the Laplace transform of the system time.

But the total arrivals during the customer's system time is simply the total number of customers in the system during his/her system time. We have derived that as given by Equation (5.16), hence we have

$$T^*(\lambda - \lambda z) = (1 - \rho)\frac{(z-1)x^*(\lambda - \lambda z)}{z - x^*(\lambda - \lambda z)} \tag{5.18}$$

After a change of variable $s = \lambda - \lambda z$, we arrive at

$$T^*(s) = (1 - \rho)\frac{sx^*(s)}{s - \lambda + \lambda x^*(s)} \tag{5.19}$$

5.2   THE RESIDUAL SERVICE TIME APPROACH

We now turn our attention to another analysis approach – Residual Service Time. In this approach, we look at the arrival epochs rather than the departure epochs and derive the waiting time of an arriving customer. According to the PASTA property, the system state seen by this arriving customer is the same as that seen by a random observer, hence the state distribution derived with respect to this arriving customer is then the limiting steady-state system distribution.

Consider the instant when a new customer (say the ith customer) arrives at the system, the waiting time in the queue of this customer should equal the sum of the service times of those customers ahead of him/her in the queue and the residual service time of the customer currently in service. The residual service time is the remaining time until service completion of the customer in service and we denote it as r_i to highlight the fact that this is the time that customer i has to wait for that customer in service to complete his/her service; as shown in Figure 5.2.

The residual service time r_i is zero if there is no customer in service when the ith customer arrives. Assume that there are n customers ahead of him/her in the waiting queue, then his/her waiting time (w_i) is given as

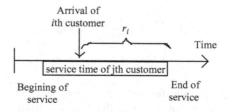

Figure 5.2   Residual service time

$$w_i = u(k)r_i + \sum_{j=i-n}^{i-1} x_j \tag{5.20}$$

where $u(k)$ is defined as follows to account for either an empty system or a system with k customers:

$$u(k) = \begin{cases} 1 & k \geq 1 \\ 0 & otherwise \end{cases} \tag{5.21}$$

If the system is assumed to be ergodic and hence a steady-state exists, taking expectation of both sides and noting that n is a random variable, we arrive at

$$E[w_i] = E[u(k)r_i] + E\left[\sum_{j=i-n}^{i-1} x_j\right] \tag{5.22}$$

Let us examine the first term on the right-hand side of the expression. Here we assume that the service time distribution is independent of the state of the system; that is the number of customers in the system. We have

$$\begin{aligned} E[u(k)r_i] &= E[u(k)]E[r_i] \\ &= \{0 \cdot P[k=0] + P[k \geq 1]\}E[r_i] \\ &= \rho R = \lambda \bar{x} R \end{aligned} \tag{5.23}$$

where $R = E[r_i]$ is the mean residual service time. The second term in Equation (5.22) is simply the random sum of independent random variables and follows the result of Example 1.7. Combining both results and Little's theorem, we obtain the following expressions:

$$\begin{aligned} E[w_i] &= \lambda \bar{x} R + E[n]E[x_j] \\ W &= \lambda \bar{x} R + \bar{x} N_q + \bar{\lambda} \bar{x} R + \bar{x} \lambda W \\ W &= \frac{\lambda \bar{x} R}{1 - \rho} \end{aligned} \tag{5.24}$$

where N_q is the expected number of customers in the waiting queue and $W = E[w_i]$.

The only unknown in Equation (5.24) is R. We note that the function of r_i is a series of triangles with the height equal to the required service time whenever a new customer starts service, and the service time decreases at a unit rate until the customer completes service, as depicted in Figure 5.3. Note that there are no gaps between those triangles because we are looking at r_i conditioned on the fact that there is at least one customer in the system and this fact is taken care of by the $u(k)$ in the expression (5.20).

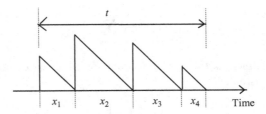

Figure 5.3 A sample pattern of the function r(t)

It is clear from Figure 5.3 that R is given by

$$R = \lim_{t \to \infty} \frac{1}{t} \int_0^t r_i(\tau) d\tau$$

$$= \lim_{t \to \infty} \frac{1}{2} \frac{\sum\limits_{k=1}^{m(t)} x_k^2}{\sum\limits_{k=1}^{m(t)} x_k} = \frac{1}{2} \lim_{t \to \infty} \frac{\frac{1}{m(t)}\sum\limits_{k=1}^{m(t)} x_k^2}{\frac{1}{m(t)}\sum\limits_{k=1}^{m(t)} x_k} = \frac{\overline{x^2}}{2\overline{x}} \qquad (5.25)$$

where $m(t)$ is the number of service completions within the time interval t.

Substituting Equation (5.25) into (5.24), we arrive at the well-known *Pollaczek-Khinchin* formula:

$$W = \frac{\lambda \overline{x^2}}{2(1-\rho)} \qquad (5.26)$$

It is interesting to note that the waiting time is a function of both the mean and the second moment of the service time. This Pollaczek-Khinchin formula is often written in terms of the coefficient of variation of the service time, C_b. Recall from Chapter 1 that C_b is the ratio of the standard deviation to the mean of a random variable (Section 1.2.4):

$$W = \frac{\rho \overline{x}}{2(1-\rho)}(1 + C_b^2) \qquad (5.27)$$

5.2.1   Performance Measures of M/G/1

Using the previously-derived result of waiting time coupled with Little's theorem, we arrive at the following performance measures:

(i)   The number of customers in the waiting queue:

$$N_q = \lambda W = \frac{\lambda^2 \overline{x^2}}{2(1-\rho)}$$ (5.28)

(ii) The time spent in the system (system time):

$$T = \overline{x} + \frac{\lambda \overline{x^2}}{2(1-\rho)}$$ (5.29)

(iii) The number of customers in the system:

$$N = \rho + \frac{\lambda^2 \overline{x^2}}{2(1-\rho)}$$ (5.30)

A special case of the M/G/1 queueing system is that of deterministic service time – the M/D/1 model. This particular queueing system is a good model for analysing packet delays of an isolated queue in a packet switching network if the packet arrival process is Poisson. This model has been used to analyse the performance of time-division multiplexing and asynchronous time-division multiplexing.

Another important application of M/G/1 queueing systems in communications networks is in the study of data link protocols, for example, Stop-and-Wait and Sliding Window protocols.

Example 5.1

Let us re-visit the train arrivals problem of Example 2.4. If the arrival process is not Poisson and the inter-train arrival times are distributed according to a general distribution with a mean of 10 minutes, when a passenger arrives at the station, how long does he/she need to wait on average until the next train arrives?

Solution

Intuitively, you may argue that since you may likely to come at the middle of an inter-arrival interval, hence the answer is 5 minutes. The actual answer is somewhat counter-intuitive to the above intuitive reasoning. Let us examine the situation more closely.

The inter-arrival intervals of the trains can be viewed as a series of service intervals experienced by customers in a queueing system and the passenger that comes to the station as an arriving customer to a queueing system. This is analogous to the situation in Figure 5.2, and the time a passenger needs to wait for the train to arrive is actually the residual service time seen by an arriving customer to a non-empty system. From the previous analysis, the mean residual service time is given by

$$R = \frac{\overline{x^2}}{2\overline{x}} = \frac{1}{2}\left(\overline{x} + \frac{\sigma_x^2}{\overline{x}}\right)$$

where σ_x^2 is the variance of the service time distribution. Hence, we can see that the average remaining time a passenger needs to wait is greater than $0.5\overline{x}$. The actual time also depends on the variance of the service time distribution.

For the Poisson process mentioned in Example 2.4, the inter-arrival times are exponentially distributed and we have

$$R = \frac{1}{2}\left(\overline{x} + \frac{\sigma_x^2}{\overline{x}}\right) = \frac{1}{2}\left(\overline{x} + \frac{(\overline{x})^2}{\overline{x}}\right) = \overline{x}$$

This is in line with the memoryless property of a Poisson process. A passenger needs to wait on average 10 minutes, regardless of when he/she comes to the station.

This situation is popularly referred to as the Paradox of Residual Life.

Example 5.2

Packets of average length L arrive at a switching node according to a Poisson process with rate λ. The single outgoing link of the switching node operates at D bps. Compare the situation where packet lengths are exponentially distributed with that where packet lengths are a fixed constant in terms of the transit time (time taken by a packet to go through the node) and the mean number of packets in the input buffer.

Solution

(i) When the packet length is fixed, the situation described can be modelled as an M/D/1 queue.
     Given the transmission time $x = L/D$, hence, $x^2 = (L/D)^2$

We have
$$T = \frac{L}{D} + \frac{\lambda(L/D)^2}{2(1-\lambda L/D)}$$

$$= \frac{L/D}{1-\lambda L/D} - \frac{\lambda(L/D)^2}{2(1-\lambda L/D)}$$

$$N = \frac{\lambda L}{D} + \frac{\lambda^2(L/D)^2}{2(1-\lambda L/D)}$$

$$= \frac{\lambda(L/D)}{1-\lambda L/D)} - \frac{\lambda^2(L/D)^2}{2(1-\lambda L/D)}$$

(ii) When the packet lengths are exponentially distributed, it is an M/M/1 model, and we have

$$T = \frac{L/D}{1-\lambda L/D}$$

$$N = \frac{\lambda(L/D)}{1-\lambda L/D}$$

We see that the constant service time case offers better performance than the exponential case as far as the transit time and number of messages in the system are concerned.

Example 5.3

In a point-to-point setup as shown in Figure 5.4, data packets generated by device A are sent over a half-duplex transmission channel operating at 64 Kbps using stop-and-wait protocol. Data packets are assumed to be generated by the device according to a Poisson process and are of fixed length of 4096 bits. The probability of a packet being in error is 0.01 and the round trip propagation delay is a constant 10 msec. Assume that Acks and Nacks are never in error.

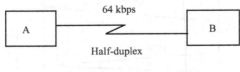

A point-to-point setup

Figure 5.4 A point-to-point setup

(i)   What is the average time required to transmit a packet until it is correctly received by B?
(ii) At what packet arrival rate will the transmission channel be saturated?
(iii) At half the arrival rate found in part (ii), what is the average waiting time of a packet before it gets transmitted?

Solution

(i)   From the data exchange sequence (Figure 5.5), it is clear that the probability that it will take exactly k attempts to transmit a packet successfully is $p^{k-1}(1-p)$, and

$$E[n] = \sum_{k=1}^{\infty} k p^{k-1}(1-p) = \frac{1}{1-p}$$

$$E[n^2] = \sum_{k=1}^{\infty} k^2 p^{k-1}(1-p) = (1-p)\sum_{k=1}^{\infty} k^2 p^{k-1}$$

$$= (1-p)\left[\frac{2}{(1-p)^3} - \frac{1}{(1-p)^2}\right] = \frac{1+p}{(1-p)^2}$$

$$T_L = transmission\ time + round\ trip\ delay$$
$$= (4096/64000) + 0.01 = 0.074\,\text{s}$$

and
$$T = E[n] \cdot T_L = \frac{1}{1-0.01} \times 0.074$$
$$= 0.07475\,\text{s}$$

(ii) Saturation occurs when the link utilization = 1, that is

$$\lambda = \mu = \frac{1}{T} = 13.38\ \text{packet/s}$$

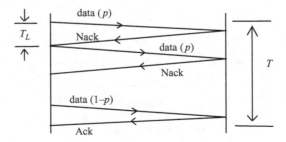

Figure 5.5 Data exchange sequence

(iii) The transmission channel can be modelled as an M/G/1 queue and the waiting time in the queue is given by the Pollaczek-Khinchin formula. Note that the rate now is $13.38/2 = 6.69$.

$$E[T^2] = T_L^2 \cdot E[n^2] = 0.005643 \, s^2$$

$$W = \frac{\lambda E[T^2]}{2(1 - \lambda T)} = 0.0378 \, s$$

5.3   M/G/1 WITH SERVICE VOCATIONS

Parallel to those extensions we had for the M/M/1 queue, we can extend this basic M/G/1 model to cater for some variation in service times. We shall look at the case where the M/G/1 server takes a rest or so-called 'vacation' after having served all the customers in the waiting queue. He/she may take another vacation if the system is still empty upon his/her return. Customers arriving during such vacation periods can go into service only after the server returns from vacation. This model could be applied in a polling type situation where a single server polls a number of stations in a pre-defined fashion. From the station's point of view, the server goes for a vacation after completing service to all customers at this station.

Assume that those successive vacations taken by the server are independent and identically distributed random variables. They are also independent of the customer inter-arrival times and service times. Using the same reasoning as before, the waiting time of a customer i in the queue before he/she receives his/her service is given by

$$W_i = u(k)r_i + u'(k)v_i + \sum_{j=i-n}^{i-1} x_j \tag{5.31}$$

where v_i is the residual vacation time; that is the remaining time to completion of a vacation when customer i arrives at the system, and $u'(k)$ is the complement of the unit-step function defined in Equation (5.21). It is defined as

$$u'(k) = \begin{cases} 0 & u(k) = 1 \\ 1 & otherwise \end{cases}$$

We use this complement function to reflect the fact that when a customer arrives at the system, he/she either sees a residual service time or falls into a vacation period, but not both. From the previous section, we have

$$E[u(k)r_i] = \rho R_s$$

hence

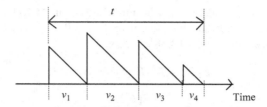

Figure 5.6   A sample pattern of the function v(t)

$$E[u'(k)v_i] = (1 - \rho)R_v.$$

Here, we denote R_s as the mean residual service time and R_v as the mean residual vacation time. As usual, taking expectation of Equation (5.31), we have

$$W = \frac{\rho}{1-\rho}R_s + R_v \qquad (5.32)$$

We already have the expression for R_s. To find R_v, we again examine the function of residual vacation time, as shown in Figure 5.6. The busy periods in between those vacation periods do not appear in the diagram because they have been taken care of by the residual service time.

Following the same arguments as before, the mean residual vacation time R_v is found to be

$$R_v = \frac{1}{2}\frac{\overline{V^2}}{\overline{V}} \qquad (5.33)$$

Therefore, substituting into the expression for W, we obtain

$$W = \frac{\lambda \overline{x^2}}{2(1-\rho)} + \frac{\overline{V^2}}{2\overline{V}} \qquad \text{or}$$

$$= \frac{\rho \overline{x}}{2(1-\rho)}(1+C_b^2) + \frac{\overline{V^2}}{2\overline{V}} \qquad (5.34)$$

5.3.1   Performance Measures of M/G/1 with Service Vacations

(i)   The system time:

$$T = \overline{x} + \frac{\lambda \overline{x^2}}{2(1-\rho)} + \frac{\overline{V^2}}{2\overline{V}} \qquad (5.35)$$

(ii) The number of customers in the system:

$$N = \rho + \frac{\lambda^2 \overline{x^2}}{2(1-\rho)} + \frac{\lambda \overline{V^2}}{2\overline{V}} \qquad (5.36)$$

(iii) The number of customers in the waiting queuing:

$$N_q = \frac{\lambda^2 \overline{x^2}}{2(1-\rho)} + \frac{\lambda \overline{V^2}}{2\overline{V}} \qquad (5.37)$$

Example 5.4

Figure 5.7 shows a multi-point computer-terminal system in which each terminal is polled in turn. When a terminal is polled, it transmits all the messages in its buffer until it is empty. Messages arrive at terminal A according to a Poisson process with a rate of 8 messages per second and the time between each poll has mean and standard deviations of 1000 ms and 250 ms, respectively. If message transmission time has the mean and standard variations of 72 ms and 10 ms, respectively, find the expected time a message arriving at A has to wait before it gets transmitted.

Solution

From the point of view of terminal A, the server (transmission channel) can be deemed as taking a vacation when other terminals are polled, thus the problem can be modelled as an M/G/1 queue with vacations:

Given: $\quad \bar{x} = 72$ ms $\qquad \sigma_{\bar{x}} = 10$ ms

$\qquad \bar{V} = 1000$ ms $\qquad \sigma_{\bar{V}} = 250$ ms

then $\qquad \overline{x^2} = \sigma^2 + (\bar{x})^2 = 10^2 + 72^2$

$\qquad \qquad = 5.284 \times 10^{-3} \text{ s}^2$

$\qquad \overline{V^2} = 1^2 + (0.025)^2 = 1.0625 \text{ s}^2$

$\qquad \lambda = 8 \quad \Rightarrow \quad \rho - 8 \times 0.072 = 0.576$

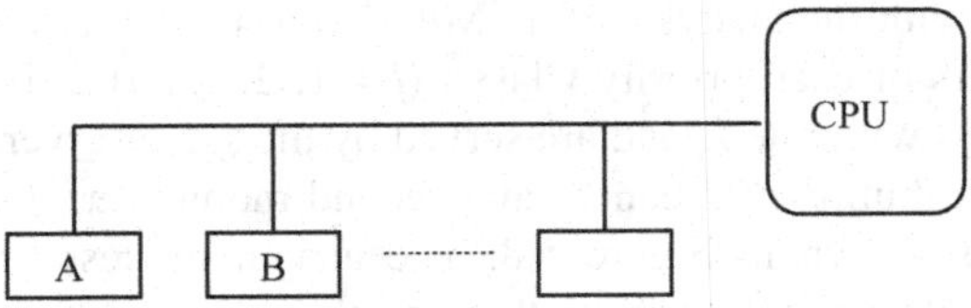

Figure 5.7 A multi-point computer terminal system

Therefore, we have

$$W = \frac{\lambda \overline{x^2}}{2(1-\rho)} + \frac{\overline{V^2}}{2\overline{V}} = 0.581\text{s}$$

5.4 PRIORITY QUEUEING SYSTEMS

For all the queueing systems that we have discussed so far, the arriving customers are treated equally and served in the order they arrive at the system; that is FCFS queueing discipline is assumed. However, in real life we often encounter situations where certain customers are more important than others. They are given greater privileges and receive their services before others. The queueing system that models this kind of situation is called a *priority queueing system*.

There are various applications of this priority model in data networks. In a packet switching network, control packets that carry vital instructions for network operations are usually transmitted with a higher priority over that of data packets. In a multi-media system in which voice and data are carried in the same network, the voice packets may again accord a higher priority than that of the data packets owing to real-time requirements.

For our subsequent treatments of priority queueing systems, we divide the arriving customers into *n* different priority classes. The smaller the priority class number, the higher the priority; i.e. Class 1 has the highest priority and Class 2 has the second highest and so on.

There are two basic queueing disciplines for priority systems, namely *pre-emptive* and *non-preemptive*. In a *pre-emptive priority queueing system*, the service of a customer is interrupted when a customer of a higher priority class arrives. As a further division, if the customer whose service was interrupted resumes service from the point of interruption once all customers of higher priority have been served, it is a *pre-emptive resume* system. If the customer repeats his/her entire service, then it is a *pre-emptive repeat* system. In the case of the non-preemptive priority system, a customer's service is never interrupted, even if a customer of higher priority arrives in the meantime.

5.4.1 M/G/1 Non-preemptive Priority Queueing

We shall begin with the analysis of an M/G/1 non-preemptive system. In this model, customers of each priority Class i ($i = 1, 2, \ldots n$) arrive according to a Poisson process with rate λ_i and are served by the same server with a general service time distribution of mean $\overline{x_i}$ and second moment $\overline{x_i^2}$ for customers of each Class i, as shown in Figure 5.8. The arrival process of each class is assumed to be independent of each other and the service process. Within each

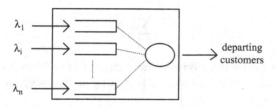

Figure 5.8   M/G/1 non-preemptive priority system

class, customers are served on their order of arrival. Again the queue for each class of customers is infinite.

If we define the total arrival $\lambda = \lambda_1 + \lambda_2 + \ldots + \lambda_n$ and utilization of each class of customers $\rho_i = \lambda \overline{x_i}$, then we have the average service time $\overline{x}$ and system utilization ρ given by

$$\overline{x} = \frac{1}{\mu} = \frac{\lambda_1}{\lambda}\overline{x_1} + \frac{\lambda_2}{\lambda}\overline{x_2} + \ldots + \frac{\lambda_n}{\lambda}\overline{x_n} \tag{5.38}$$

$$\rho = \frac{\lambda}{\mu} = \rho_1 + \rho_2 + \ldots + \rho_n \tag{5.39}$$

The system will reach equilibrium if $\Sigma \rho_i < 1$. However, if this condition is violated then at least some priority classes will not reach equilibrium.

Now let us look at a 'typical' customer C_n of Class i who arrives at the system. His/her mean waiting time in the queue is made up of the following four components:

(i)   The mean residual service time R for all customers in the system.
When a Class i customer arrives, the probability that he/she finds a Class j customer in service is $\rho_j = \lambda_j \overline{x_j}$, therefore R is given by the weighted sum of the residual service time of each class, as shown in Equation (5.40). Note that the term in the brackets is the residual service time for Class k, as found in Equation (5.25):

$$R = \sum_{k=1}^{n} \rho_k \left(\frac{\overline{x_k^2}}{2\overline{x_k}} \right) = \frac{1}{2}\sum_{k=1}^{n} \lambda_k \overline{x_k^2} \tag{5.40}$$

(ii) The mean total service time of those customers of the same class ahead of him/her in the waiting queue, that is $\overline{x_i}N_q^i$, where N_q^i is the average number of customers of Class i in the waiting queue.

(iii) The mean total service time of those customers of Class $j (j < i)$ found in the system at the time of arrival; i.e.:

$$\sum_{j=1}^{i-1} \overline{x_j} N_q^j.$$

(iv) The mean total service time of those customers of Class $j (j < i)$ arriving at the system while customer C_n is waiting in the queue, i.e.:

$$\sum_{j=1}^{i-1} \overline{x_j} \lambda_j W_i$$

Combining the four components together, we arrive at

$$W_i = R + \overline{x_i} N_q^i + \sum_{j=1}^{i-1} \overline{x_j} N_q^j + \sum_{j=1}^{i-1} \overline{x_j} \lambda_j W_i \qquad (5.41)$$

For Class 1 customers, since their waiting times are not affected by customers of lower classes, the expression of W_1 is the same as that of an M/G/1 queue and is given by

$$W_1 = \frac{R}{1 - \rho_1} \qquad (5.42)$$

By using Equation (5.41) together with Equation (5.42) and the expression $N_q^k = \lambda_k W_k$, we can obtain the mean waiting time for Class 2 customers as

$$W_2 = \frac{R}{(1 - \rho_1)(1 - \rho_1 - \rho_2)} \qquad (5.43)$$

In general, the expression for the mean waiting time for Class i customers can be calculated recursively using the preceding approach and it yields

$$W_i = \frac{R}{(1 - \rho_1 - \rho_2 - \dots \rho_{i-1})(1 - \rho_1 - \rho_2 - \dots \rho_i)} \qquad (5.44)$$

where the mean residual service time R is given by expression (5.40).

5.4.2 Performance Measures of Non-preemptive Priority

The other performance measures can be found once the waiting time is known:

(i) The average number of customers of each class in their own waiting queue:

$$(N_q)_i = \lambda_i W_i = \frac{\lambda_i R}{(1 - \rho_1 - \ldots - \rho_{i-1})(1 - \rho_1 - \ldots - \rho_i)} \tag{5.45}$$

(ii) The total time spends in the system by a customer of Class i:

$$T_i = W_i + \overline{x_i} \tag{5.46}$$

(iii) Total number of customers in the system:

$$N = \sum_{k=1}^{n}(N_q)_k + \rho \tag{5.47}$$

(iv) If the service times of each class of customers are exponentially distributed with a mean of μ, then in effect we have an M/M/1 non-preemptive priority system. Then we have

$$R = \frac{1}{2}\sum_{k=1}^{n}\lambda_k\left(\frac{1}{\mu}\right)^2 = \frac{\rho}{\mu} \tag{5.48}$$

Note that the variance and mean of an exponential distribution with parameter μ are μ^{-2} and μ^{-1}, respectively, hence the second moment is $2\mu^{-2}$.

Example 5.5

In a packet switching network, there are two types of packets traversing the network; namely data packets of fixed length of 4800 bits and control packets of fixed length of 192 bits. On average, there are 15% control packets and 85% data packets present in the network and the combined arriving stream of packets to a switching node is Poisson with rate $\lambda = 4$ packets per second. If all transmission links in the network operate at 19.2 kbps, calculate:

(i) the average waiting time for both the control and data packets at a switching node.
(ii) the waiting times for the control and data packets respectively if a non-preemptive priority scheme is employed at each switching node with control packets given a higher priority.

Solution

Given: Data packets $\quad \overline{x_d} = \dfrac{4800}{19200} = 0.25, \quad \sigma_d = 0 \quad$ and

$$\overline{x_d^2} = (0.25)^2 = 0.0625$$

Control packets $\quad \overline{x_c} = \dfrac{192}{19200} = 0.01, \quad \sigma_c = 0 \quad$ and $\quad \overline{x_c^2} = 0.0001$

and $\qquad\qquad\qquad \lambda_d = 0.85\lambda \quad$ and $\quad \lambda_c = 0.15\lambda$

(i) Without the priority scheme, the combined traffic stream can be modelled as an M/G/1 queue with the second moment of the composite stream given by the weighted sum of their individual second moments:

$$\overline{x^2} = \frac{\lambda_c}{\lambda}\overline{x_c^2} + \frac{\lambda_d}{\lambda}\overline{x_d^2}$$
$$= 0.05314$$

and

$$\overline{x^2} = \frac{\lambda_c}{\lambda}\overline{x_c} + \frac{\lambda_d}{\lambda}\overline{x_d}$$
$$= 0.214$$

Therefore $\qquad W = \dfrac{\lambda \overline{x^2}}{2(1-\rho)} = \dfrac{4 \times 0.05314}{2(1 - 4 \times 0.214)}$
$$= 0.738\,\mathrm{s}$$

(ii) With priority incorporated, we have

$$\rho_c = 0.15\lambda \overline{x_c} = 0.006$$
$$\rho_d = 0.85\lambda \overline{x_d} = 0.85$$
$$R = \frac{1}{2}\sum \lambda_i \overline{x_i^2} = 0.10628$$

Therefore:

$$W_c = \frac{R}{1-\rho_c} = \frac{0.10628}{1 - 0.006}$$
$$= 0.10692\,\mathrm{s}$$

$$W_d = \frac{R}{(1-\rho_c)(1-\rho_c-\rho_d)} = \frac{0.10628}{0.994 \times 0.144}$$

$$= 0.747 \, s$$

We see from this example that the priority scheme reduces the waiting time of control packets substantially without increasing greatly the waiting time of data packets.

Example 5.6

Consider an M/G/1 queue whose arrivals comprise of two classes of customers (1 and 2) with equal arrival rates. Class 1 customers have a higher priority than Class 2 customers. If all the customers have exponential distributed service times with rate μ, calculate the length of the waiting queue for various values of ρ.

Solution

Given $\lambda_1 = \lambda_2 = 0.5\lambda$, we have

$$(N_q)_1 = \frac{\rho^2}{4-2\rho} \quad \text{and} \quad (N_q)_2 = \frac{\rho^2}{(4-2\rho)(1-\rho)}$$

therefore

ρ	$(N_q)_1$	$(N_q)_2$
0.6	0.129	0.321
0.7	0.188	0.628
0.8	0.267	1.333
0.9	0.368	3.682
0.95	0.43	8.667
0.97	0.457	15.2

From the above table, we see that at higher and higher utilization factors, more and more Class 1 customers are served earlier at the expense of Class 2 customers. The queue of Class 2 customers grows rapidly while that of Class 1 customers stays almost constant.

5.4.3   M/G/1 Pre-emptive Resume Priority Queueing

This model is the same as the previous one except that now a customer in service can be pre-empted by customers of higher priority. The interrupted service resumes from the point of interruption when all customers of higher priority classes have been served. There are two basic distinctions between this mode and the previous one:

(i)   The presence of customers of lower priority Classes ($i + 1$ to n) in the system has no effect on the waiting time of a customer of Class i because he/she always pre-empts those customers of lower priority classes. So in the analysis, we can ignore those customers of Class $i + 1$ to n.

(ii) While a customer of Class i is waiting for his/her service, his/her waiting time is the same whether customers of Class 1 to $i - 1$ are served in a pre-emptive manner or non-preemptive fashion. This is due to the fact that he/she only gets his/her service when there are no customers of higher priority classes in the system.

      We can use the expression (5.44) for his/her waiting time. Thus, the waiting time in queue of a customer of Class i is given by

$$W_i = \frac{R_i}{(1 - \rho_1 - \ldots - \rho_{i-1})(1 - \rho_1 - \ldots - \rho_i)} \quad \text{and} \quad R_i = \frac{1}{2}\sum_{k=1}^{i} \lambda_k \overline{x_k^2} \quad (5.49)$$

However, the system time of a customer of Class i is not equal to W_i plus his/her service time because his/her service may be interrupted by customers of higher Classes (1 to $i - 1$) arriving while he/she is being served. Let us define T_k' to be the time his/her service starts until completion, then we have the following expression:

$$T_i' = \overline{x_i} + \sum_{j=1}^{i-1} (\overline{x_j}) \lambda_j T_i' \quad (5.50)$$

where $\lambda_j T_i'$ is the average arrival of customers of Class j ($j = 1$ to $i - 1$) during the time T_i'. Combining these two parts, the system time of a customer of Class i is then given by

$$T_i = W_i + T_i'$$

$$T_i = \frac{R_i}{(1 - \rho_i - \ldots - \rho_{i-1})(1 - \rho_1 - \ldots - \rho_i)} + \frac{\overline{x_i}}{(1 - \rho_1 - \ldots - \rho_{i-1})} \quad (5.51)$$

and we arrive at the following expression:

$$T_i = \frac{\overline{x_i}(1 - \rho_1 - \ldots - \rho_1) + R_i}{(1 - \rho_1 - \ldots - \rho_{i-1})(1 - \rho_1 - \ldots - \rho_i)} \quad i > 1 \quad (5.52)$$

For $i = 1$, we have

$$T_1 = \frac{\overline{x_1}(1 - \rho_1) + R_1}{1 - \rho_1} \tag{5.53}$$

Example 5.7

Consider Example 5.5, we repeat the calculation assuming the pre-emptive resume priority scheme is employed.

Solution

$$R_1 = \frac{1}{2}\lambda_c \overline{x_c^2} = 0.00003$$

$$R_2 = \frac{1}{2}(\lambda_c \overline{x_c^2} + \lambda_d \overline{x_d^2}) = 0.10628$$

$$T_c = \frac{\overline{x_c}(1 - \rho_c) + R_1}{1 - \rho_c} = 0.01003\,s$$

$$T_d = \frac{\overline{x_d}(1 - \rho_c - \rho_d) + R_2}{(1 - \rho_c)(1 - \rho_c - \rho_d)} = 0.994\,s$$

Compare them with that of the non-preemptive scheme:

$$T_c = W_c + \overline{x_c} = 0.11692\,s$$
$$T_d = W_d + \overline{x_d} = 0.9925\,s$$

Again, we see that the pre-emptive scheme significantly reduces the system time of the control packets without increasing too much the system time of those data packets.

5.5   THE G/M/1 QUEUEING SYSTEM

The G/M/1 queueing system can be considered as the 'dual image' of M/G/1. This model is less useful than M/G/1 in data networks, owing to the reasons given in Section 4.8. However, it is important theoretically and worth mentioning because this is the first time (in so many chapters) that we see that the system state seen by an arriving customer is different from that seen by a random observer.

In this model, customers arrive according to a general arrival process with mean rate λ and are served by an exponential server with mean μ^{-1}. The inter-arrival times of those customers are assumed to be independent and identically distributed random variables. Some literature uses GI instead of G to signify this nature of independence.

The stochastic process $\{N(t)\}$ is not Markovian because the elapsed time since the last arrival has to be considered in deriving P_k. But if we focus our attention at those instants where an arrival occurs, the process $\{N(t)|An\ arrival\}$ at those arrival instants can be shown to be a Markovian process and so solved by the imbedded Markov-chain technique.

We will not discuss the analysis of this model as it is rather complex, but just state the results. It has been shown (Kleinrock 1975) that the probability of finding k customers in the system immediately before an arrival is given by the following:

$$R_k = (1-\sigma)\sigma^k \quad k = 0, 1, 2, \ldots \tag{5.54}$$

where σ is the unique root of the equation:

$$\sigma = A^*(\mu - \mu\sigma) \tag{5.55}$$

and $A^*(\mu - \mu\sigma)$ denotes the Laplace transform of the *pdf* of inter-arrival times evaluated at the special point $(\mu - \mu\sigma)$.

As discussed earlier, R_k is, in general, not equal to P_k. They are equal only when the arrival process is Poisson. It can be shown that P_k is given by

$$P_k = \begin{cases} \rho R_{k-1} & k = 1, 2, 3, \ldots \\ 1-\rho & k = 0 \end{cases} \tag{5.56}$$

5.5.1   Performance Measures of GI/M/1

(i)   The number of customers in the waiting queue:

$$N_q = \sum_{k=0}^{\infty} k P_{k+1} + \sum_{k=1}^{\infty} k\rho(1-\sigma)\sigma^k$$

$$= \frac{\rho\sigma}{1-\sigma} \tag{5.57}$$

(ii)   The number of customers in the system:

$$N = \sum_{k=0}^{\infty} k P_k = \frac{\rho}{1-\sigma} \tag{5.58}$$

(iii) The waiting time in the queue:

$$W = \frac{N_q}{\lambda} = \frac{\sigma}{\mu(1-\sigma)} \qquad (5.59)$$

(iv) The time spent in the system:

$$T = \frac{N}{\lambda} = \frac{1}{\mu(1-\sigma)} \qquad (5.60)$$

Problems

1. Consider a switch with two input links and one outgoing transmission link. Data packets arrive at the first input link according to a Poisson process with mean λ_1 and voice packets arrive at the second input link also according to a Poisson process with mean λ_2. Determine the total transit time when a packet arrives at either input until its transmission completion if the service time of both the data and voice packets are exponentially distributed with mean rates μ_1 and μ_2, respectively.

2. Consider a time-division multiplexor that multiplexes 30 input packet streams onto a outgoing link with a slot time of 4 μs. Assume that the packet arrival process of each packet stream is Poisson with a rate of 3000 packets/sec. What is the average waiting time before a packet is transmitted if the outgoing link transmits a packet from each input line and then instantaneously switches to serve the next input line according to a fixed service cycle?

3. Two types of packets, naming Control and Data packets, arrive at a switching node as independent Poisson processes with a rate of 10 packets/sec (Control packets) and 30 packets/sec (Data packets), respectively. Control packets have a constant length of 80 bits and are given higher priority over Data packets for processing. Data packets are of exponential length with a mean of 1000 bits. If the only outgoing link of this node is operating at 64 000 bps, determine the mean waiting times of these two types of packets for a non-preemptive priority system. Repeat the calculation when the priorities of these two types of packets are switched.

4. Repeat Question 3 if the pre-emptive priority system is used.

5. Consider a queueing system where customers arrive according to a Poisson process with rate λ but the service facility now consists of two servers in series. A customer upon entry into the service facility will spend a random amount of time with server 1, and then proceeds

immediately to the second server for service after leaving server 1. While the customer in the service facility is receiving his/her service from either server, no other customer is allowed into the service facility. If the service rates of these two servers are exponentially distributed with rate 2μ, calculate the mean waiting time of a customer in this queueing system.

6. Consider the queueing system in Problem 5, but the service facility now consists of two parallel servers instead of serial servers. A customer upon entry into the service facility will proceed to either server 1 with probability 0.25 or to server 2 with probability 0.75. While the customer in the service facility is receiving his/her service, no other customer is allowed into the service facility. If the service rates of these two servers are exponentially distributed with rate $\mu_i(i = 1,2)$, calculate the mean waiting time of a customer in this queueing system.

6

Open Queueing Networks

In preceding discussions, we have dealt solely with single isolated queueing systems and showed examples of their applications in data networks. It is a natural extension to now look at a collection of interactive queueing systems, the so-called networks of queues, whereby the departures of some queues feed into other queues. In fact, queueing networks are a more realistic model for a system with many resources interacting with each other. For simplicity, each individual queueing system is often referred to as a queue in a queueing network or just a node. We will use this terminology when the context is clear.

The analysis of a queueing network is much more complicated due to the interactions between various queues and we have to examine them as a whole. The state of one queue is generally dependent on the others because of feedback loops and hence the localized analysis of an isolated queue will not give us a complete picture of the network dynamics.

There are many real-life applications that can be modelled as networks of queues. In communication networks, those cascaded store-and-forward switching nodes that forward data packets/messages from one node to another are a good example of a network of queues. Job processing on a machine floor is another candidate for the model of network queues. A job usually requires a sequence of operations by one or more machines. A job entering the machine floor corresponds to an arrival, and its departure occurs when its required processing at various machines has been completed.

There are various ways of classifying queueing networks. From the network topology point of view, queueing networks can be categorized into two generic classes, namely Open and Closed Queueing Networks, and of course a mixture

Queueing Modelling Fundamentals Second Edition Ng Chee-Hock and Soong Boon-Hee
© 2008 John Wiley & Sons, Ltd

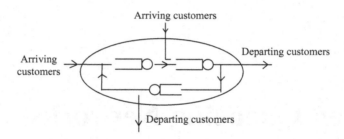

Figure 6.1   An example of open queueing networks

of the two. Alternatively, queueing networks can be classified according to the queue capacity at each queuing node. In a queueing network, where all queues have infinite capacity then we have the so-called Non-Blocking Networks (or Queueing Networks without Blocking). On the other hand, if one or more queues are of finite capacity, resulting in customers being blocked when the waiting queue is full, then we have the blocking networks (or Queueing Networks with Blocking).

There may be multiple classes of customers traversing a network. A multi-class network will have a number of classes of customers with different arrival/service patterns while traversing a network on different paths.

In this chapter, we use the commonly used classification of open and closed queueing networks, and we will only look at the networks with a single class of customers:

(i) Open queueing networks

An open queueing network is one where customers arrive from external sources outside the domain of interest, go through several queues or even revisit a particular queue more than once, and finally leave the system, as depicted in Figure 6.1. Inside the network, a customer finishing service at any queue may choose to join a particular queue in a deterministic fashion or proceed to any other queues probabilistically using a pre-defined probability distribution. Note that the total sum of arrival rates entering the system is equal to the total departure rate under steady-state condition – *flow conservation principle*.

In addition, an open queueing network is a feed-forward network if customers visit queues within the network in an acyclic fashion without re-visiting previous queues – no feedback paths.

Open queueing networks are good models for analysing circuit-switching and packet-switching data networks. However, there are some complications involved. Certain simplifying assumptions have to be adopted in applying these models to data networks (see Section 6.2).

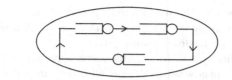

Figure 6.2 An example of closed queueing networks

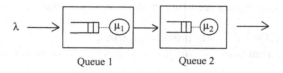

Queue 1   Queue 2

Figure 6.3 Markovian queues in tandem

(ii) Closed queueing networks

In a closed queueing network, customers do not arrive at or depart from the system. There are a constant number of customers simply circulating through the various queues and they may revisit a particular queue more than once, as in the case of open queueing networks. Again, a customer finishing service at one queue may go to another queue deterministically or probabilistically. A sample of closed queueing networks is shown in Figure 6.2.

Closed queueing networks may appear to be unusual and unrealistic, but they are good models for analysing window-type network flow controls as well as CPU job scheduling problems. In a CPU job scheduling problem, there is a large number of jobs waiting to be scheduled at all times, yet the number of jobs served in the system is fixed at some value and a job enters the system immediately whenever the service of another job is completed. This situation is typically modelled as a closed queueing network.

Instead of plunging head-on into the analysis of a general open queueing network, we shall look at a very simple class of open queueing networks – queues in series. To begin with, we will first consider the two queues in tandem. The result will be extended to more general cases in Burke's theorem.

6.1 MARKOVIAN QUEUES IN TANDEM

This is the situation where two queues are joined in series, as shown in Figure 6.3. Customers arrive at Queue 1 according to a Poisson process of mean λ and are served by an exponential server with mean μ_1^{-1}. After completion of service, customers join the second queue and are again served by an exponential server, who is independent of the server of Queue 1, with mean μ_2^{-1}. As usual, we assume that the arrival process is independent of any internal processes in both queues. The waiting time at both queues are assumed to be infinite so that no blocking occurs. This is an example of a simple feedback open network where no feedback path exists.

The analysis of this system follows the same approach as that of a single Markovian queue except now we are dealing with a two-dimensional state space. Let us focus our attention on the system state (k_1, k_2), where k_1 and k_2 are the numbers of customers in Queue 1 and Queue 2, respectively. Since the arrival process is Poisson and the service time distributions are exponential, we can have only the following events occurring in an incremental time interval Δt:

- an arrival at Queue 1 with probability $\lambda \Delta t$;
- a departure from Queue 1, and hence an arrival at Queue 2, with probability $\mu_1 \Delta t$;
- a departure from Queue 2 with probability $\mu_2 \Delta t$;
- no change in the system state with probability $[1 - (\lambda + \mu_1 + \mu_2)\Delta t]$.

Using the same technique that we employed in studying birth-death processes, by considering the change in the joint probability $P(k_1, k_2)$ in an infinitesimal period of time Δt, we have

$$P(k_1, k_2; t + \Delta t) = P(k_1 - 1, k_2; t)\lambda \Delta t + P(k_1 + 1, k_2 - 1; t)\mu_1 \Delta t$$
$$+ P(k_1, k_2 + 1; t)\mu_2 \Delta t + P(k_1, k_2; t)[1 - (\lambda + \mu_1 + \mu_2)\Delta t] \quad (6.1)$$

Rearranging terms and dividing the equation by Δt and letting Δt go to zero, we arrive at a differential equation for the joint probability:

$$P'(k_1, k_2; t) = P(k_1 - 1, k_2; t)\lambda + P(k_1 + 1, k_2 - 1; t)\mu_1$$
$$+ P(k_1, k_2 + 1; t)\mu_2 - P(k_1, k_2; t)[\lambda + \mu_1 + \mu_2] \quad (6.2)$$

where $P'(k_1, k_2; t)$ denotes the derivative. If $\lambda < \mu_1$ and $\lambda < \mu_2$, the tandem queues will reach equilibrium. We can then obtain the steady-state solution by setting the differentials to zero:

$$(\mu_1 + \mu_2 + \lambda)P(k_1, k_2) = \mu_1 P(k_1 + 1, k_2 - 1) + \mu_2 P(k_1, k_2 + 1)$$
$$+ \lambda P(k_1 - 1, k_2) \quad (6.3)$$

Using the same arguments and repeating the process, we obtain three other equations for the boundary conditions:

$$\lambda P(0, 0) = \mu_2 P(0, 1) \quad (6.4)$$

$$(\mu_2 + \lambda)P(0, k_2) = \mu_1 P(1, k_2 - 1) + \mu_2 P(0, k_2 + 1) \quad k_2 > 0 \quad (6.5)$$

$$(\mu_1 + \lambda)P(k_1, 0) = \mu_2 P(k_1, 1) + \lambda P(k_1 - 1, 0) \quad k_1 > 0 \quad (6.6)$$

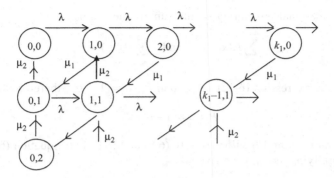

Figure 6.4 State transition diagram of the tandem queues

Students will be quick to notice that this set of equations resembles that of the one-dimensional birth-death processes and can be interpreted as flow balancing equations for probability flow going in and out of a particular state. The state transition diagram which reflects this set of flow equations is shown in Figure 6.4.

Similar to the one-dimensional case, this set of equations is not independent of each other and the additional equation required for a unique solution is provided by the normalization condition:

$$\sum_{k_1} \sum_{k_2} P(k_1, k_2) = 1 \tag{6.7}$$

To solve Equations (6.3) to (6.6), let us assume that the solution has the so-called *Product form*, that is

$$P(k_1, k_2) = P_1(k_1)P_2(k_2) \tag{6.8}$$

where $P_1(k_1)$ is the marginal probability function of Queue 1 and is a function of parameters of Queue 1 alone; similarly $P_2(k_2)$ is the marginal probability function of Queue 2 and is a function of parameters of Queue 2 alone.

Substituting Equation (6.8) into Equation (6.4), we have

$$\lambda P_2(0) = \mu_2 P_2(1) \tag{6.9}$$

Using it together with Equation (6.8) in Equation (6.6), we have

$$\mu_1 P_1(k_1) = \lambda P_1(k_1 - 1) \tag{6.10}$$

$$P_1(k_1) = \frac{\lambda}{\mu_1} P_1(k_1 - 1)$$
$$= \rho_1^{k_1} P_1(0) \quad \text{where} \quad \rho_1 = \frac{\lambda}{\mu_1} \tag{6.11}$$

Since the marginal probability should sum to one, we have

$$\sum_{k_1} P_1(k_1) = 1 \quad \text{and} \quad \sum_{k_2} P_2(k_2) = 1 \tag{6.12}$$

Using the expression (6.12), we obtain $P_1(0) = (1 - \rho_1)$, therefore

$$P_1(k_1) = (1 - \rho_1)\rho_1^{k_1} \tag{6.13}$$

Now, substituting Equations (6.10), (6.8) and (6.9) into Equation (6.3) and after simplifying terms, we arrive at

$$(\lambda + \mu_2)P_2(k_2) = \lambda P_2(k_2 - 1) + P_2(k_2 + 1)\mu_2 \tag{6.14}$$

This is a recursive equation in $P_2(k_2)$, which we solve by z-transform:

$$\sum_{k_2=1}^{\infty} P_2(k_2)(\lambda + \mu_2)z^{k_2} = \sum_{k_2=1}^{\infty} \lambda P_2(k_2 - 1)z^{k_2} + \sum_{k_2=1}^{\infty} P_2(k_2 + 1)\mu_2 z^{k_2} \tag{6.15}$$

Define the z-transform of $P_2(k_2)$ as

$$P_2(z) = \sum_{k_2=0}^{\infty} P_2(k_2)z^{k_2} \tag{6.16}$$

We have from Equation (6.15):

$$(\lambda + \mu)[P_2(z) - P_2(0)] = \lambda z P_2(z) + \frac{\mu_2}{z}\left[P_2(z) - P_2(0) - \frac{\lambda}{\mu_2}P_2(0) \right]$$

$$P_2(z)(\rho_2 z - 1)(z - 1) = P_2(0)(1 + \rho_2)(1 - z) \quad \text{where} \quad \rho_2 = \lambda/\mu_2$$

and

$$P_2(z) = P_2(0)\frac{1 + \rho_2}{1 - \rho_2 z} \tag{6.17}$$

Similarly, since the marginal probability P_2 should sum to 1, that is equivalent to $P_2(z)|_{z=1} = 1$, we arrive at

$$P_2(0) = \frac{1 - \rho_2}{1 + \rho_2} \tag{6.18}$$

and

$$P_2(z) = (1 - \rho_2)\frac{1}{1 - \rho_2 z} \tag{6.19}$$

Inverting the z-transform, we have

$$P_2(k_2) = (1 - \rho_2)\rho_2^{k_2} \tag{6.20}$$

Therefore:

$$P(k_1, k_2) = (1 - \rho_1)\rho_1^{k_1}(1 - \rho_2)\rho_2^{k_2} \tag{6.21}$$

where $\rho_1 = \dfrac{\lambda}{\mu_1}$ and $\rho_2 = \dfrac{\lambda}{\mu_2}$

The expression of (6.21) holds for $\rho_0 < 1$ and $\rho_1 < 1$.

For an isolated M/M/1 queueing system, the probability that there are k customers in the system is $P(k) = (1 - \rho)\rho^k$, therefore

$$\begin{aligned} P(k_1, k_2) &= (1 - \rho_1)\rho_1^{k_1}(1 - \rho_2)\rho_2^{k_2} \\ &= P_1(k_1)P_2(k_2) \end{aligned}$$

We see that the joint probability distribution is the product of the marginal probability distributions, and hence the term *Product-Form* solution.

6.1.1   Analysis of Tandem Queues

The foregoing analysis provides us with a steady-state solution but fails to give us an insight into the interaction between the two queues. The final expression seems to suggest that the two queues are independent of each other. Are they really so?

To answer that, let us examine the two tandem queues in more detail. First let us look at Queue 1. The customer arriving pattern to Queue 1 is a Poisson process and the service times are distributed exponentially, therefore it is a classical M/M/1 queue.

How about Queue 2, what is the customer arriving pattern, or in other words the inter-arrival time distribution? It is clear from the connection diagram that the inter-departure time distribution from Queue 1 forms the inter-arrival time distribution of Queue 2. It can be shown that the customer arriving pattern to Queue 2 is a Poisson process as follows.

When a customer, say A, departs from Queue 1, he/she may leave behind an empty system with probability $(1 - \rho_1)$ or a busy system with probability ρ_1. In the case of an empty system, the inter-departure time between A and the next customer (say B) is the sum of B's service time and the inter-arrival time between A and B at Queue 1. Whereas in the busy system case, the inter-departure time is simply the service time of B.

Therefore, the Laplace transform of the density function for the unconditional inter-departure time (I) between A and B is given by

$$L[f_I(t)] = (1-\rho_1)\left[\frac{\lambda}{s+\lambda}\cdot\frac{\mu_1}{s+\mu_1}\right] + \rho_1\frac{\mu_1}{s+\mu_1}$$

$$= \frac{\lambda}{s+\lambda} \tag{6.22}$$

This is simply the Laplace transform of an exponential density function. Hence
a Poisson process driving an exponential server generates a Poisson departure
process. Queue 2 can be modelled as an M/M/1 queue.

6.1.2   Burke's Theorem

The above discussion is the essence of Burke's theorem. In fact, Burke's
theorem provides a more general result for the departure process of an M/M/m
queue instead of just the M/M/1 queue discussed earlier. This theorem states
that the steady-state output of a stable M/M/m queue with input parameter λ
and service-time parameter μ for each of the m servers is in fact a Poisson
process at the same rate λ. The output is independent of the other processes in
the system.

Burke's theorem is very useful as it enables us to do a queue-by-queue
decomposition and analyse each queue separately when multiple-server queues
(each with exponential *pdf* service-times) are connected together in a feed
forward fashion without any feedback path.

Example 6.1

Sentosa Island is a famous tourist attraction in Singapore. During peak hours,
tourists arrive at the island at a mean rate of 35 per hour and can be approxi-
mated by a Poisson process. As tourists complete their sightseeing, they queue
up at the exit point to purchase tickets for one of the following modes of trans-
portation to return to the mainland; namely cable car, ferry and mini-bus. The
average service time at the ticket counters is 5 minutes per tourist.

Past records show that a tourist usually spends an average of 8 hours on
sightseeing. If we assume that the sightseeing times and ticket-purchasing
times are exponentially distributed; find:

i) the minimum number of ticket counters required in operation during peak
    periods.

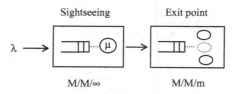

Figure 6.5 Queueing model for example 6-1

ii) If it is decided to add one more than the minimum number of counters required in operation, what is the average waiting time for purchasing a ticket? How many people, on average, are on the Sentosa Island?

Solution

We note that since each tourist roams about freely on the island, each tourist is a server for himself, hence the portion where each tourist goes around the island can be modelled as an M/M/∞ queue. The checkout counters at the exit is the M/M/m queue. Hence, the overall model for the problem is as shown in Figure 6.5.

The first queue has the following parameters:

$$\lambda = 35 \text{ /hr} \quad \text{and} \quad \mu_1 = 8 \text{ hr}$$

The input to the second queue is the output of the first queue. From Burke's theorem, we know that the output of the first queue is Poisson, hence the input to second queue is also Poisson, with $\lambda = 35$ /hr:

(i) To have a stable situation for the second queue:

$$\rho = \frac{\lambda}{m\mu} < 1.$$

Given $1/\mu_2 = 5$ min, we have

$$m > \lambda/\mu_2 = 35/(60/5) = 2.92$$

hence, 3 counters are required:

(ii) For 4 counters, we have an M/M/4 queue with $\lambda = 35$ and $1/\mu_2 = 5$ min

$$P_0 = \left[\sum_{k=0}^{m-1} \frac{(m\rho)^k}{k!} + \frac{(m\rho)^m}{m!(1-\rho)} \right]^{-1}$$

$$= \left[\sum_{k=0}^{3} \frac{1}{k!} \left(\frac{35}{12} \right)^k + \frac{(35/12)^4}{4!\left(1 - \frac{35}{48}\right)} \right]^{-1} = 0.0427$$

$$W = \frac{P_d}{m\mu_2 - \lambda} = \frac{1}{m\mu_2 - \lambda} \times \frac{P_0(m\rho)^m}{m!(1-\rho)}$$

$$= \frac{1}{13} \cdot \frac{(35/12)^4}{4!(13/48)} P_0$$

$$= 0.036\,hr = 2.16\,\text{min}$$

and the average number of people waiting in line to purchase tickets N_q:

$$N_q = \lambda W$$
$$= 35 \times (0.036)$$
$$= 1.26$$

$$N_2 = N_q + \frac{\lambda}{\mu_2}$$
$$= 1.99$$

The total number of tourists (N) on the island is the sum of the number of tourists (N_1) doing their sightseeing and the number of tourists (N_2) at the exit point. It can be shown that the number of customers in an M/M/∞ is given by

$$N_1 = \frac{\lambda}{\mu_1} = \frac{35}{(1/8)}$$
$$= 280$$

Therefore $N = N_1 + N_2$
$$= 282$$

6.2   APPLICATIONS OF TANDEM QUEUES IN DATA NETWORKS

In a data network, messages (or packets) usually traverse from node to node across several links before they reach their destination. Superficially, this picture gives us an impression that the transmission of data packets across several links can in a way be modelled as a cascade of queues in series. In reality, certain simplifying assumptions need to be made before the queueing results can be applied.

To understand the complications involved, we will consider two transmission channels of same transmission speed in tandem in a packet switching network. This is a situation which is very similar to the queue in tandem that we discussed earlier. Unfortunately it warrants further consideration owing to the dependency between the two queues, as illustrated below:

(i)   If packets have equal length, then the first transmission channel can be modelled as an M/D/1 queue whereas the second transmission channel cannot be modelled as an M/D/1 queue because the departure process of the first queue is no longer Poisson.

(ii) If packet lengths are exponentially distributed and are independent of each other as well as the inter-arrival times at the first queue, the first queue can be modelled as an M/M/1 queue but the second queue still cannot be modelled as an M/M/1 because the inter-arrival times are strongly correlated with the packet lengths and hence the service times. A packet that has a transmission time of t seconds at the first channel will have the same transmission time at the second.

The second queue can be modelled as an M/M/1 only if the correlation is not present, i.e. a packet will assume a new exponential length upon departure from a transmission channel. Kleinrock (1975) suggested that merging several packet streams on a transmission channel has an effect similar to restoring the independence of inter-arrival times and packet lengths. This assumption is known as *Kleinrock independence approximation*.

With the Kleinrock's approximation suggestion and coupled with Burke's theorem, we are now in a position to estimate the average end-to-end delay of acyclic store-and-forward data networks. Virtual circuits in a packet switching network is a good example of a feed-forward queues in which packets traverse a virtual circuit from source to destination, as shown in Figure 6.6.

If we focus on a single virtual circuit then it generally has no feedback loop and can be approximately modelled as a series of queues, as shown in Figure 6.7.

The situation will not be so simple and straightforward when there are feedback loops. The presence of feedback loops destroys the Poisson characteristic of the flow and Burke's theorem will no longer be valid. We shall look at this type of network when we discuss Jackson's queueing networks.

Referring to Figure 6.7, if we assume that packets of exponentially distributed lengths arrive at Queue 1 according to a Poisson process and further assume that they take on new exponential lengths while traversing a cascade of queues in series, then the virtual circuit path can be decomposed into several M/M/1 queues and the average end-to-end delay for packets is then given by

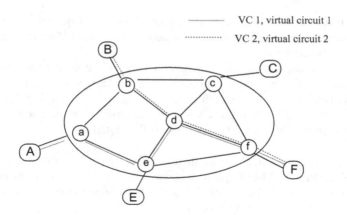

Figure 6.6   A virtual circuit packet switching network

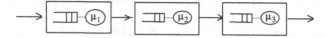

Figure 6.7   Queueing model for a virtual circuit

$$E(T) = \sum_{i=1}^{k} \frac{1}{\mu_i - \lambda_i} \qquad (6.23)$$

where μ_i and λ_i are the service rate and arrival rate of Queue i respectively. k is the number of links in that virtual circuit and hence the number of queues in series.

Example 6.2

To illustrate the application, let us look at Figure 6.6. Assume that all links are operating at 19.2 kbps and there are only two virtual circuit paths, VC1 and VC2, being setup at the moment. The traffic pattern is as shown below:

Data Flow	Data Rate (pks/s)	mean packet length	path
B to F	10	800 bits	VC2
A to F	5	800 bits	VC1
E to F	3	800 bits	VC1

Calculate the end-to-end delays for:

(i)   packets sent from station B to station F along VC2;
(ii)   packets sent from station A to station F along VC1;
(iiii) packets sent from station E to station F along VC1.

Solution

Using Kleinrock independence approximation and assuming Poission packet stream, we have

$$\mu = \frac{19200}{800} = 24 \text{ packets/s}$$

(i) $\quad E(T_{bdf}) = \dfrac{1}{24-10} + \dfrac{1}{24-18} = \dfrac{5}{21} \text{ s}$

(ii) $\quad E(T_{aedf}) = \dfrac{1}{24-5} + \dfrac{1}{24-8} + \dfrac{1}{24-18} = \dfrac{257}{912} \text{ s}$

(iii) $\quad E(T_{edf}) = \dfrac{1}{24-8} + \dfrac{1}{24-18} = \dfrac{11}{48} \text{ s}$

6.3   JACKSON QUEUEING NETWORKS

Having studied the tandem queues, we shall now look at more general open queueing networks. We shall examine only a class of open queueing networks – Jackson open queueing networks which exhibit product-form solutions and hence lend themselves to mathematical tractability.

A Jackson queueing network is a network of an M M/M/m state-independent queueing system (hereafter referred as a queueing node or simply node), as shown in Figure 6.8, with the following features:

- There is only one class of customers in the network.
- Customers from the exterior arrive at each node i according to a Poisson process with rate $\gamma_i \geq 0$.
- All customers belong to the same class and their service times at node i are all exponentially distributed with mean μ_i. The service times are independent from that at other nodes and are also independent of the arrival process.
- A customer upon receiving his/her service at node i will proceed to node j with a probability p_{ij} or leave the network at node i with probability:

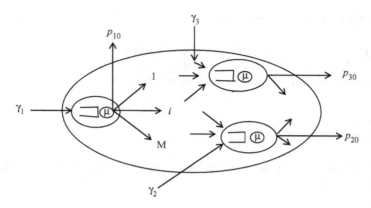

Figure 6.8   An open queueing network

$$p_{io} = 1 - \sum_{j=1}^{M} p_{ij}.$$

- The queue capacity at each node is infinite so there is no blocking.

If we assume the network reaches equilibrium, we can then write the following two traffic equations using the flow conservation principle. The first equation is for a particular node in the network. It shows that the total sum of arrival rates from other nodes and that from outside the domain of interest to a particular node is equal to the departure rate from that node. The second expression is for the network as a whole. It equates the total arrival rates from outside the domain of interest to the total departure rates from the network. It should be noted that we are assuming only the existence of a steady state and not Poisson processes in these two equations. The arrival process to each individual queue is generally not Poisson:

$$\lambda_i = \gamma_i + \sum_{j=1}^{M} \lambda_j p_{ji} \quad j = 1, \ldots, M \tag{6.24}$$

$$\sum_{i=1}^{M} \gamma_i = \sum_{i=1}^{M} \lambda_i p_{io}$$

$$= \sum_{i=1}^{M} \lambda_i \left(1 - \sum_{j=1}^{M} p_{ij} \right) \tag{6.25}$$

Here, λ_i is the effective arrival rate to node i. Putting the flow equations in matrix form, we have

$$\lambda_i - \sum_{j=1}^{M} \lambda_j p_{ji} = \gamma_i \quad i = 1, 2, \ldots, M \tag{6.26}$$

$$\tilde{I}\tilde{\lambda} - \tilde{P}^T \tilde{\lambda} = \tilde{\gamma}$$
$$(\tilde{I} - \tilde{P}^T)\tilde{\lambda} = \tilde{\gamma} \tag{6.27}$$

where

$\tilde{I}$ is the identity matrix

$$\tilde{\lambda}^T = [\lambda_1, \lambda_2, \ldots, \lambda_M]$$

$$\tilde{\gamma}^T = [\gamma_1, \gamma_2, \ldots, \gamma_M]$$

$$\tilde{P} = \begin{bmatrix} p_{11} & p_{12} & \cdots & p_{1M} \\ p_{21} & p_{22} & \cdots & \\ \cdots & \cdots & \cdots & \cdots \\ p_{M1} & & & p_{MM} \end{bmatrix} \tag{6.28}$$

Jackson demonstrated in 1957 (Jackson 1957) that the joint probability distribution for this type of network exhibits the product-form solution. i.e.:

$$P(\tilde{n}) = P(n_1, n_2, \ldots, n_M)$$
$$= P_1(n_1)P_2(n_2) \ldots P_M(n_M) \tag{6.29}$$

where $P_i(n_i)$ is the marginal probability distribution of each node i.

We present here a simple verification of the product-form solution assuming that all nodes are in an M/M/1 queue. In fact, the Jackson's result is applicable to multi-server and state dependent cases. To begin with, we define the following notations:

$$\tilde{n} = (n_1, n_2, \ldots, n_i, \ldots, n_M)$$

$$\tilde{1}_i = (0, 0, \ldots, 1, 0, 0)$$

$$\tilde{n} - \tilde{1}_i = (n_1, n_2, \ldots, n_i - 1, \ldots, n_M)$$

$$\tilde{n} + \tilde{1}_i = (n_1, n_2, \ldots, n_i + 1, \ldots, n_M)$$

Parallel to the analysis of tandem queues, we again focus our attention on the change in system state $\tilde{n}$ in an incremental time interval Δt. The system as a whole is Markov because the arrival processes are all Poisson and service processes are exponential, therefore there can only be four types of transition, as follows:

- no change in system state
- an arrival to node i
- a departure from node i
- transfer of a customer from node i to node j.

The corresponding transitional probability for the network going from state $\tilde{m}$ to state $\tilde{n}$ due to the above mentioned four events is

$$P(\tilde{n}|\tilde{m}) = \begin{cases} 1 - \sum_{i=1}^{M} \gamma_i \Delta t - \sum_{i=1}^{M} \mu_i \Delta t & \tilde{m} = \tilde{n} \\ \gamma_i \Delta t & \tilde{m} = \tilde{n} + \tilde{1}_i \\ \mu_i \left(1 - \sum_{j=1}^{M} p_{ij} \right) \Delta t & \tilde{m} = \tilde{n} - \tilde{1}_i \\ \mu_i p_{ij} \Delta t & \tilde{m} = \tilde{n} + \tilde{1}_j - \tilde{1}_i \end{cases}$$

Therefore

$$P[\tilde{n}(t + \Delta t)] = P[\tilde{n}(t)] \left(1 - \sum_{i=1}^{M} \gamma_i \Delta t - \sum_{i=1}^{M} \mu_i \Delta t \right)$$
$$+ \sum_{i=1}^{M} P[\tilde{n}(t) - \tilde{1}_i] \gamma_i \Delta t + \sum_{i=1}^{M} P[\tilde{n}(t) + \tilde{1}_i] \mu_i \left(1 - \sum_{j=1}^{M} p_{ij} \right) \Delta t$$
$$+ \sum_{i=1}^{M} \sum_{j=1}^{M} P[\tilde{n}(t) + \tilde{1}_j - \tilde{1}_i] \mu_j p_{ji} \Delta t \qquad (6.30)$$

Rearranging terms, dividing the equation by Δt, taking limit and dropping the time parameter, we have

$$\frac{d}{dt} P(\tilde{n}) = P(\tilde{n}) \left(-\sum_{i=1}^{M} \gamma_i - \sum_{i=1}^{M} \mu_i \right) + \sum_{i=1}^{M} \gamma_i P(\tilde{n} - \tilde{1}_i)$$
$$+ \sum_{i=1}^{M} P(\tilde{n} + \tilde{1}_i) \mu_i \left(1 - \sum_{j=1}^{M} p_{ij} \right) + \sum_{i=1}^{M} \sum_{j=1}^{M} \mu_j p_{ji} P(\tilde{n} + \tilde{1}_j - \tilde{1}_i) \quad (6.31)$$

Setting the derivative of the joint probability density function to zero, we obtain

$$\left[\sum_{i=1}^{M} \gamma_i + \sum_{i=1}^{M} \mu_i \right] P(\tilde{n}) = \sum_{i=1}^{M} \gamma_i P(\tilde{n} - \tilde{1}_i) + \sum_{i=1}^{M} \mu_i \left(1 - \sum_{j=1}^{M} p_{ij} \right) P(\tilde{n} + \tilde{1}_i)$$
$$+ \sum_{i=1}^{M} \sum_{j=1}^{M} \mu_j p_{ji} P(\tilde{n} - \tilde{1}_i + \tilde{1}_j) \qquad (6.32)$$

This is a rather complex equation to deal with, so instead of solving the equation directly we do the reverse. With a bit of hindsight, let us suggest that the final solution $P(\tilde{n})$ satisfies the following relationships:

$$P(\tilde{n} - \tilde{1}_i) = \rho_i^{-1} P(\tilde{n}) \tag{6.33}$$

$$P(\tilde{n} + \tilde{1}_i) = \rho_i P(\tilde{n}) \tag{6.34}$$

$$P(\tilde{n} - \tilde{1}_i + \tilde{1}_j) = \rho_i^{-1} \rho_j P(\tilde{n}) \tag{6.35}$$

where as usual $\rho_k = \lambda_k / \mu_k$. The validity of this suggestion can be verified by showing that they indeed satisfy the general flow balance Equation (6.32). Substituting them into Equation (6.32), we have

$$\sum_{i=1}^{M} \gamma_i + \sum_{i=1}^{M} \mu_i = \sum_{i=1}^{M} \gamma_i \rho_i^{-1} + \sum_{i=1}^{M} \mu_i \left(1 - \sum_{j=1}^{M} p_{ij} \right) \rho_i + \sum_{i=1}^{M} \sum_{j=1}^{M} \mu_j p_{ji} \rho_i^{-1} \rho_j$$

So far we have not used the flow conservation Equations (6.24) and (6.25). The above equation can be further simplified by using these two equations. In doing so, we arrive at

$$\sum_{i=1}^{M} \gamma_i + \sum_{i=1}^{M} \mu_i = \sum_{i=1}^{M} \left(\lambda_i - \sum_{j=1}^{M} \lambda_j p_{ji} \right) \rho_i^{-1} + \sum_{i=1}^{M} \gamma_i + \sum_{i=1}^{M} \sum_{j=1}^{M} \lambda_j p_{ji} \rho_i^{-1}$$

$$= \sum_{i=1}^{M} \mu_i + \sum_{i=1}^{M} \gamma_i \tag{6.36}$$

Now, we have established the validity of our suggested relationships of $P(\tilde{n})$, it is easy for us to find the final expression of $P(\tilde{n})$ using expression (6.33):

$$P(\tilde{n}) = \rho_i P(\tilde{n} - \tilde{1}_i)$$

$$= \rho_i P(n_1, n_2, \ldots, n_i - 1, \ldots, n_M)$$

$$= \rho_i^{n_i} P(n_1, n_2, \ldots, 0, \ldots, n_M)$$

$$= P(0, 0, \ldots, 0) \prod_{i=1}^{M} \rho_i^{n_i}$$

$$= P(\tilde{0}) \prod_{i=1}^{M} \rho_i^{n_i} \tag{6.37}$$

The $P(\tilde{0})$ can be found, as usual, by the normalization condition:

$$\sum_{\tilde{n}} P(\tilde{n}) = P(\tilde{0}) \sum_{\tilde{n}} \left[\prod_{i=1}^{M} \rho_i^{n_i} \right] = 1$$

$$P(\tilde{0}) \prod_{i=1}^{M} \sum_{n_i=0}^{\infty} \rho_i^{n_i} = 1$$

$$P(\tilde{0}) \prod_{i=1}^{M} (1 - \rho_i)^{-1} = 1$$

$$P(\tilde{0}) = \prod_{i=1}^{M} (1 - \rho_i) \tag{6.38}$$

Hence, the final solution is given by

$$P(\tilde{n}) = \prod_{i=1}^{M} (1 - \rho_i) \rho_i^{n_i}$$

$$= P_1(n_1) P_2(n_2) \ldots P_M(n_M) \tag{6.39}$$

The condition for stability in the network is $\lambda_i/\mu_i < 1$ and we shall assume that this condition exists throughout all our discussions.

There are two observations we can derive from the about expression. Firstly, the expression seems to suggest that each queue in the network behaves as if it is independent of the others and hence the joint state probability is a product of the marginal probabilities. Secondly, each queue can be considered separately in isolation, even though the arrival is not Poisson due to the feedback paths.

6.3.1   Performance Measures for Open Networks

Jackson's theorem coupled with Little's theorem provides us with a simple means of evaluating some of the performance parameters of an open queueing network:

(i)   Total throughput of the network γ.
      This follows from the flow conservation principle, as the total number of customers entering the system should be the same as the total number of customers leaving the system, if it is stable.

$$\gamma = \sum_{i} \gamma_i \tag{6.40}$$

(ii)   The average number of times a customer visits a node V_i is

$$V_i = \lambda_i/\gamma \qquad (6.41)$$

and the network-wide average number of times a customer visits a typical node V is

$$V = \sum_i \lambda_i \Big/ \gamma \qquad (6.42)$$

(iii)   The marginal probability of node i is, in general, defined as

$$P_i(k) = \sum_{n_i=k} P(n_1, n_2, \ldots, n_M) \qquad (6.43)$$

Subject to the normalization condition

$$\sum P(\tilde{n}) = 1.$$

(iv)   Once the marginal probability is obtained, the marginal queue length for the ith node in theory can be calculated. For a product-form open queueing network consisting of only single-server queues, the marginal queue length is the same as that of an isolated M/M/1, as shown below. The arrival rate λ_i (or load) to each node can be obtained from the overall traffic pattern using flow conservation Equations (6.24) and (6.25):

$$n_i = \frac{\rho_i}{1-\rho_i} \quad \text{and} \quad \rho_i = \frac{\lambda_i}{\mu_i} \qquad (6.44)$$

Then the total number of customers in the whole system (network) is

$$N = n_1 + n_2 + \ldots + n_M$$
$$= \sum_{i=1}^{M} \frac{\rho_i}{1-\rho_i}$$
$$= \sum_{i=1}^{M} \frac{\lambda_i}{\mu_i - \lambda_i} \quad \text{(all M/M/1 queues)} \qquad (6.45)$$

(v)   Average time spent in the network:

$$T = \frac{N}{\sum_{i=1}^{M} \gamma_i} = \frac{1}{\sum_{i=1}^{M} \gamma_i} \sum_{i=1}^{M} \frac{\lambda_i}{\mu_i - \lambda_i} \qquad (6.46)$$

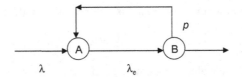

Figure 6.9   A queueing model for Example 6.3

Example 6.3

Consider the situation where a station A transmits messages in the form of packets to another station B via a single link operating at 64 k bps. If B receives a packet with an error, A will re-transmit that packet until it is correctly received. By assuming that packets of exponential length with a mean of 1 k bytes are generated at A at a rate of 4 packets per second, and the probability that a packet transmitted and arrived at B with transmission errors is p = 0.1, calculate the utilization of this link.

Solution

The situation can be modelled as shown in Figure 6.9 with the re-transmission as the feedback path:
The effective arrival of packets to the link is

$$\lambda_e = \lambda + p\lambda_e \quad \Rightarrow \quad \lambda_e = \frac{\lambda}{1-p}$$

Since

$$\lambda = 4 \quad \text{and} \quad \mu = \frac{64000}{1000 \times 8} = 8$$

The link utilization is then

$$\rho = \frac{\lambda_e}{\mu} = \frac{\lambda}{(1-p)\mu} = \frac{4}{0.9 \times 8} = 0.56$$

Example 6.4

Figure 6.10 shows a queueing model of a multi-programming computer system. Computer jobs arrive at the CPU according to a Poisson process with rate γ. A

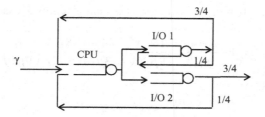

Figure 6.10 A multi-programming computer

job gets executed in the CPU for an exponential time with mean μ_1^{-1} and then requests service from either I/O 1 or I/O 2 with equal probability. The processing times of these two I/O devices are exponentially distributed with means μ_2^{-1} and μ_3^{-1}, respectively. After going through either one of the I/O devices, the job may return to the CPU or leave the system according to the probabilities shown on the diagram. Calculate:

(i)   the joint probability mass function of the whole network;
(ii) if $\mu_1 = 8\gamma$ and $\mu_2 = \mu_3 = 4\gamma$, find the mean number of jobs at each queue;
(iii) the mean number of jobs in the network, and the time a job spends in the network.

Solution

Let λ_1, λ_2, λ_3 be the effective arrival rates to the CPU , I/O 1 and I/O 2, respectively. By the flow conservation principle, we have

$$\lambda_1 = \gamma + \frac{3}{4}\lambda_2 + \frac{1}{4}\lambda_3$$

$$\lambda_2 = \frac{1}{2}\lambda_1 + \frac{1}{4}\lambda_2$$

$$\lambda_3 = \frac{1}{2}\lambda_1$$

Solving them, we obtain

$$\lambda_1 = \frac{8}{3}\gamma, \quad \lambda_2 = \frac{16}{9}\gamma, \quad \lambda_3 = \frac{4}{3}\gamma$$

(i)   The joint probability mass function is then given by Jackson's theorem:

$$P(n_1, n_2, n_3) = \left(1 - \frac{8\gamma}{3\mu_1}\right)\left(1 - \frac{16\gamma}{9\mu_2}\right)\left(1 - \frac{4\gamma}{3\mu_3}\right)\left(\frac{8\gamma}{3\mu_1}\right)^{n_1}\left(\frac{16\gamma}{9\mu_2}\right)^{n_2}\left(\frac{4\gamma}{3\mu_3}\right)^{n_3}$$

(ii) $\rho_1 = \dfrac{\lambda_1}{\mu_1} = \dfrac{1}{3}, \quad \rho_2 = \dfrac{4}{9}, \quad \rho_3 = \dfrac{1}{3}$

$N_{CPU} = \dfrac{\rho_1}{1 - \rho_1} = \dfrac{1/3}{1 - (1/3)} = \dfrac{1}{2}$

$N_{I/O1} = \dfrac{4}{5} \quad \text{and} \quad N_{I/O2} = \dfrac{1}{2}$

(iii) $N = N_{CPu} + N_{I/O1} + N_{I/O2} = \dfrac{9}{5}$

$T = \dfrac{9/5}{\gamma} = \dfrac{9}{5\gamma}$

6.3.2   Balance Equations

Students who are familiar with the probability flow balance concept discussed earlier will be able to recognize that Equation (6.32) is the network version of the global balance equation encountered in Chapter 4. The equation basically equates the total probability flow in and out of the state $\tilde{n}$.

The left-hand side of the equation is the probability flow leaving state $\tilde{n}$ due to arrivals or departure from the queues. The first term on the right corresponds to the probability flow from state $(\tilde{n} - \tilde{1}_i)$ into state $\tilde{n}$ due to an arrival, the second term is the probability flow from state $(\tilde{n} + \tilde{1}_i)$ into state $\tilde{n}$ due to a departure and the last term is the probability flow from $(\tilde{n} - \tilde{1}_i + \tilde{1}_j)$ into state $\tilde{n}$ due to a transfer of a customer from node i to j.

The expressions (6.33) to (6.35) are simply the local balance equations. These local balance equations must satisfy the global balance equation. It should be noted that the local balance concept is a unique property peculiar to product-form queueing networks. It is a stronger and more stringent form of balance than global balance. A solution that satisfies all the local balance equations also satisfies the global balance equations, but the converse does not necessary hold.

Example 6.5

Consider the virtual-circuit packet-switching network shown in Figure 6.6, and the traffic load together with the routing information as given in Table 6.1. If

Table 6.1 Traffic load and routing information

	a	b	c	e	f
a	—	4 (ab)	1 (abc)	3 (ae)	6 (aef)
b	5 (ba)	—	4 (bc)	1 (bde)	10 (bdf)
c	2 (cba)	2 (cb)	—	1 (cde)	3 (cf)
e	1 (ea)	3 (eab)	7 (edc)	—	1 (ef)
f	10 (fea)	1 (fdb)	2 (fc)	4 (fe)	—

*The number in each cell indicates the mean number of packets per second that are transmitted from a source (row) to a destination (column). The alphabet string indicates the path taken by these packets.

all the links operate at 9600 bps and the average length of a packet is 320 bits, calculate:

(i) The network-wide packet delay;
(ii) The average number of hops a packet will traverse through the network.

Solution

(i) The network-wide delay is given by the expression

$$T = \frac{1}{\displaystyle\sum_{i=1}^{M} \gamma_i} \sum_{i=1}^{M} \frac{\lambda_i}{\mu_i - \lambda_i}$$

Let us calculate the last summation using the data listed in Table 6.2. Since the transmission speed is the same for all links:

$$\mu_i = 9600/320 = 30$$

The total load to each link (λ_i) can be obtained from Table 6.1. For example, Link 1 (ab) has a total load of 4(ab) + 1(abc) + 3(eab) = 8 and Link 4 (ea) has a total traffic load of 1 (ea) + 3 (eab) + 10 (fea) = 14.
Once:

$$\sum_{i=1}^{M} \frac{\lambda_i}{\mu_i - \lambda_i} = 5.5499$$

is calculated, it remains for us to calculate the sum of total external arrival rates to the network $\displaystyle\sum_{i=1}^{M} \gamma_i$. This can be obtained by assuming packets

Table 6.2   Traffic load and transmission speeds

Link Number	VC Path	Load (λ_i)	$\dfrac{\lambda_i}{\mu_i - \lambda_i}$
1	ab	8	0.3636
2	ba	7	0.3044
3	ae	9	0.4286
4	ea	14	0.8750
5	bc	5	0.2000
6	cb	4	0.1538
7	bd	11	0.5789
8	db	1	0.0345
9	cd	1	0.0345
10	dc	7	0.3044
11	de	2	0.0714
12	ed	7	0.3044
13	ef	7	0.3044
14	fe	14	0.8750
15	df	10	0.5000
16	fd	1	0.0345
17	cf	3	0.1111
18	fc	2	0.0714
Total		113	5.5499

originated from those nodes a, b, c, e and f come from the external sources A, B, C, E and F. Hence, the sum of the entries in each row of Table 6.1 represents the total arrival rate from the external source attached to that node. For example, the sum of the first row $4 + 1 + 3 + 6 = 14$ represents the total arrival rate from node a. The sum of the second row $5 + 4 + 1 + 10 = 20$ represents the total arrival rate from node b, and so on.

Therefore $\sum_{i=1}^{M} \gamma_i$ can be calculated from Table 6.1 by summing the number in each cell and is equal to 71. We have

$$T = \frac{1}{\displaystyle\sum_{i=1}^{M} \gamma_i} \sum_{i=1}^{M} \frac{\lambda_i}{\mu_i - \lambda_i} = \frac{5.5499}{71}$$

$$= 0.078\,\text{s}$$

(ii) The number of hops experienced by a packet

$$V = \frac{\displaystyle\sum_i \lambda_i}{\displaystyle\sum_i \gamma_i} = \frac{113}{71} = 1.592$$

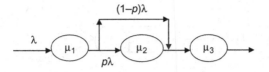

Figure 6.11 An open network of three queues

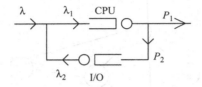

Figure 6.12 CPU job scheduling system

Problems

1. Consider an open queueing network of three queues, as shown in Figure 6.11. The output of Queue 1 is split by a probability p into two streams; one goes into Queue 2 and the other goes into Queue 3 together with the output of Queue 2. If the arrival to Queue 1 is a Poisson process with rate λ and the service times at all queues are exponentially distributed with the rates shown in the diagram, find the mass function of the state probability for the whole network.

2. Consider a closed queueing network of two queues where the departure of Queue 1 goes into Queue 2 and that of Queue 2 goes into Queue 1. If there are 4 customers circulating in the network and the service times at both queues are exponentially distributed with rates μ_1 and μ_2, respectively:

   (i) draw the state transition diagram taking the state of the network as (k_1, k_2), where $k_i (i = 1, 2)$ is the number of customers at queue i;
   (ii) write down the global balance equations for this state transition diagram.

3. Consider a CPU job-processing system in which jobs arrive at the scheduler according to a Poisson process with rate λ. A job gets executed in the CPU for an exponential time duration with mean $1/\mu_1$ and it may leave the system with probability P_1 after receiving service at CPU. It may request I/O processing with a probability P_2 and return back to CPU later. The I/O processing time is also exponentially distributed with a mean $1/\mu_2$. The queueing model of this system is depicted in Figure 6.12:

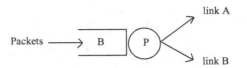

Figure 6.13   A schematic diagram of a switching node

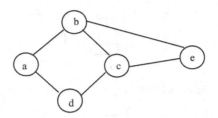

Figure 6.14   A 5-node message switching

(i) Compute the joint probability mass function for the system.
(ii) Compute the number of jobs in the whole system.

4. Figure 6.13 shows a schematic diagram of a node in a packet switching network. Packets which are exponentially distributed arrive at the
   big buffer B according to a Poisson process and gain access to the
   node processor P according to the FCFS (First-come First-served)
   service discipline. Processor P is a switching processor which directs
   one-third of the traffic to the outgoing link A (3 channels) and two-
   thirds to link B (3 channels). The node P processing time can be
   assumed to be exponentially distributed with a mean processing time
   of 5 msec/packet.

Assuming that:

- the output buffers are never saturated;
- the mean arrival rate of packets to buffer B is 160 packet/
  second;
- the mean packets' length is 800 bits; and
- the transmission speed of all channels is 64 kbits/sec.

estimate the mean transit time across the node (time between arrival
at the node and completion of transmission on an outgoing channel) for
messages routed:

(i) over link A
(ii) over link B.

Table 6.3 Traffic load and routing information for Figure 6.14

	a	b	c	d	e
a	—	5 (ab)	7 (adc)	1 (ab)	3 (abc)
b	2 (ba)	—	3 (bc)	1 (bcd)	5 (bc)
c	3 (cba)	4 (cb)	—	4 (cd)	10 (ce)
d	10 (da)	5 (dcb)	3 (dc)	—	1 (dce)
e	5 (ecda)	2 (eb)	4 (ec)	1 (ecd)	—

5. Consider the message-switching network shown in Figure 6.14, and the traffic load together with the routing information as shown in Table 6.3. If all links operate at 19.2 kbps and the average length of a message is 960 bits, calculate:

   (i) the network-wide message delay; and
   (ii) the average delay experience by a message traversing the path adce.

7

Closed Queueing Networks

In this chapter, we shall complete our analysis of queueing networks by looking at another class that has neither external arrivals nor departures – closed queueing networks. As before, we shall restrict our discussion to where all customers in the networks have the same service demand distribution. These types of networks are often termed as single customer class (or single class) networks in contrast to multi-class networks in which customers belong to different classes and each class has its own service demand distribution. Students should not confuse these multi-class networks with priority queueing systems. In multi-class queueing networks, customers of different classes are treated equally and selected for service according to the prevailing queueing discipline, even though they have different service-time distributions.

Closed queueing networks are often more useful than their open counterparts because the infinite customer population assumption that is implicit in open queueing networks is unrealistic. Many interesting problems in computer and communication networks can be formulated as closed queueing networks.

7.1 JACKSON CLOSED QUEUEING NETWORKS

For simplicity, we shall deal only with a class of closed queueing networks that parallel the Jackson open queueing networks. The structure of a Jackson closed network is similar to that of the open network. It is a closed queueing network of M queues with N customers circulating in the network with the following characteristics:

- All customers have the same service demand distribution. They are served in their order of arrivals at node i and their service demands are exponentially distributed with an average rate μ_i, which is independent of other processes in the network.
- A customer completing service at node i will proceed to node j with a probability of p_{ij}.
- There is no blocking at each node, in other words, the capacity of each node is greater than or equal to $(N - 1)$.

The equilibrium state distribution is given by the joint probability vector:

$$P(\tilde{n}) \equiv P(n_1, n_2, \ldots, n_M)$$

Since there are no external arrivals or departures in a closed queueing network, by setting $p_{io} = 0$ and $\gamma_i = 0$ in Equation (6.24), we obtain the following traffic equations:

$$\sum_{i=1}^{M} P_{ij} = 1 \tag{7.1}$$

$$\lambda_i = \sum_{j=1}^{M} \lambda_j p_{ji} \tag{7.2}$$

Equation (7.2) is a set of homogeneous linear equations that has no unique solution. Any solution $\tilde{e} = (e_1, e_2, \ldots, e_M)$ or its multiplier $k\tilde{e}$ is also a solution. Here we use the new notation e_i instead of λ_i because λ_i is no longer the absolute arrival rate to Queue i and the ratio (λ_i/μ_i) is also not the actual utilization (ρ_i) of Queue i.

To solve this equation, we can set one of the e_i s to a certain convenient value, such as 1. If we set $e_1 = 1$, then $e_i(i > 1)$ can be interpreted as the mean number of visits by a customer to node $i(i > 1)$ relative to the number of visits to station 1, which is also 1. Therefore, e_i is sometimes called the relative visitation rate of Queue i.

The state space (S) of a closed queueing network is finite since there are a fixed number of customers circulating in the system. The state space (S) is

$$S = \left\{ (n_1, n_2, \ldots n_M) \middle| n_i \geq 0 \ \& \ \sum_{i=1}^{M} n_i = N \right\} \tag{7.3}$$

Using the method shown in Example 1.2, the number of states in S will be

$$|S| = \binom{N+M-1}{M-1} = \binom{N+M-1}{N} \tag{7.4}$$

7.2   STEADY-STATE PROBABILITY DISTRIBUTION

In 1967, Gordon and Newell showed that this type of closed queueing network exhibits a product-form solution as follows:

$$P(\tilde{n}) = \frac{1}{G_M(N)} \prod_{i=1}^{M} \left(\frac{e_i}{\mu_i} \right)^{n_i} \tag{7.5}$$

where $G_M(N)$ is a normalization constant and n_i is the number of customers at node i. The above expression can be intuitively verified by the following arguments.

As there are no external arrivals or departures from the system, by setting $\gamma_i = 0$ and $p_{io} = 1 - \sum_{j=1}^{M} p_{ij}$ to zero in the expression (6.32), we have

$$\sum_{i=1}^{M} \mu_i P(\tilde{n}) = \sum_{i=1}^{M} \sum_{j=1}^{M} \mu_j p_{ji} P(\tilde{n} + \tilde{1}_j - \tilde{1}_i) \tag{7.6}$$

The expression so obtained is the *global balance equation*. The product-form solution of Equation (7.5) can therefore be derived by solving this global balance equation together with the traffic Equation (7.2). However, it is a difficult task. From the foregoing discussions on open queueing networks, it may be reasonable to suggest the following local balance equation as before:

$$P(\tilde{n}) = \left(\frac{e_i}{\mu_i} \right) \left(\frac{e_j}{\mu_j} \right)^{-1} P(\tilde{n} + \tilde{1}_j - \tilde{1}_i) \quad i, j = 1, 2, \dots, M \tag{7.7}$$

Again the validity of this suggestion can be proved by substituting it in expression (7.6) together with the traffic equation. The expression (7.7) is just one of the local balance Equations (6.33) to (6.35) seen in Section 6.3. Hence, it is equivalent to Equation (6.33) which can be used to obtain the final result:

$$P(\tilde{n}) = P(\tilde{0}) \prod_{i=1}^{M} \left(\frac{e_i}{\mu_i} \right)^{n_i} = \frac{1}{G_M(N)} \prod_{i=1}^{M} \left(\frac{e_i}{\mu_i} \right)^{n_i} \tag{7.8}$$

The normalization constant $G_M(N)$ is used in the expression instead of $P(\tilde{0})$ to signify that it can be evaluated recursively, as we shall show in the next section. By definition, it is given by the usual normalization of probability $\sum_{\tilde{n} \subset S} P(\tilde{n}) = 1$, i.e.:

$$G_M(N) = \sum_{\tilde{n} \subset S} \left\{ \prod_{i=1}^{M} \left(\frac{e_i}{\mu_i} \right)^{n_i} \right\} \qquad (7.9)$$

It is clear from the above that $G_M(N)$ is a function of both the number of queues (M) and the total number of customers (N) in the network. Students should also realize that it is also a function of e_i's, μ_i's and p_{ij}'s, although they are not explicitly expressed in the notation.

Once the normalization constant is calculated, we can then calculate the performance measures. Since there is neither external arrival nor departure from the network, certain performance measures for opening networks are not applicable here, such as the throughput of the network.

The marginal probability of node i is the probability that node i contains exactly $n_i = k$ customers. It is the sum of the probabilities of all possible states that satisfies $\sum_{i=1}^{M} n_i = N$ and $n_i = k$; i.e.:

$$P_i(k) = \sum_{\sum n_i = N \ \& \ n_i = k} P(n_1, n_2, \ldots, n_M) \qquad (7.10)$$

Again, subject to the normalization condition:

$$\sum P(\tilde{n}) = 1.$$

Interestingly, we shall see in the next section that most of the statistical parameters of interest for a closed queueing network can be obtained in terms of $G_M(N)$ without the explicit evaluation of $P(\tilde{n})$.

Example 7.1

Consider a closed queueing network, as shown in Figure 7.1, with two customers circulating in the network. Customers leaving Queue 2 and Queue 3 will automatically join back to Queue 1 for service. If service times at the three queues are exponentially distributed with rate μ and the branching probability at the output of Queue 1 is $p = 0.5$, find the joint probability distribution and other performance measures for the network.

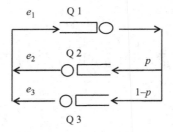

Figure 7.1 A closed network of three parallel queues

Solution

Using the flow balance at the branching point, we have

$$e_2 = e_3 = 0.5e_1$$

hence

$$P(\tilde{n}) = \frac{1}{G_M(n)} \left(\frac{e_1}{\mu}\right)^{n_1} \left(\frac{e_2}{\mu}\right)^{n_2} \left(\frac{e_3}{\mu}\right)^{n_3}$$

$$= \frac{1}{G_M(n)} \left(\frac{e_1}{\mu}\right)^{n_1+n_2+n_3} \left(\frac{1}{2}\right)^{n_2+n_3}$$

Here n_i is the number of customers at Queue I, but

$$n_1 + n_2 + n_3 = 2$$

therefore

$$P(\tilde{n}) = \frac{1}{G_M(n)} \left(\frac{e_1}{\mu}\right)^2 \left(\frac{1}{2}\right)^{n_2+n_3} = \frac{1}{G_M(n)} \left(\frac{e_1}{\mu}\right)^2 \left(\frac{1}{2}\right)^{2-n_1}$$

Now, it remains for us to calculate the normalization constant. The six possible states for the network are

$$(0,0,2),\quad (0,1,1),\quad (0,2,0)\quad (1,0,1),\quad (1,1,0),\quad (2,0,0).$$

Hence

$$G_M(N) = \sum_{\tilde{n} \subset S} \left\{ \prod_{i=1}^{M} \left(\frac{e_i}{\mu_i} \right)^{n_i} \right\} = \left(\frac{e_1}{\mu} \right)^2 \left\{ \left(\frac{1}{2} \right)^2 + \left(\frac{1}{2} \right)^2 + \left(\frac{1}{2} \right)^2 + \frac{1}{2} + \frac{1}{2} + 1 \right\}$$

$$= \frac{11}{4} \left(\frac{e_1}{\mu} \right)^2$$

Therefore, we have

$$P(\tilde{n}) = \frac{4}{11} \left(\frac{1}{2} \right)^{2-n_1}$$

The utilization of Queue 1 can be calculated as follows. Firstly, let us calculate the probability that Queue 1 is idle:

$$P[Q_1 \text{ is idle}] = P(0, 0, 2) + P(0, 1, 1) + P(0, 2, 0)$$

$$= \frac{3}{11}$$

Hence, the utilization of Queue 1 $= 1 - P[Q1 \text{ is idle}] = \frac{8}{11}$

Example 7.2

Again consider the closed queueing network shown in Figure 7.1. Let us investigate the underlying Markov chain for this queueing network. For the six possible states (n_1, n_2, n_3) mentioned in the previous example, the state transition diagram is shown in Figure 7.2.

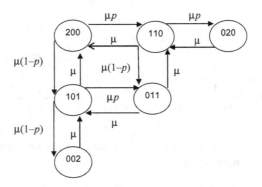

Figure 7.2   Transition diagram for Example 7.2

Writing the global balance equation for each state using the notation of $P(n_1$ n_2 $n_3)$ for state probability, we have

(200) $\mu P(200) = \mu P(110) + \mu P(101)$
(110) $2\mu P(110) = \mu P(020) + \mu P(011) + 0.5\,\mu P(200)$
(020) $\mu P(020) = 0.5\,\mu P(110)$
(101) $2\,\mu P(101) = 0.5\,\mu P(200) + \mu P(011) + \mu P(002)$
(011) $2\,\mu P(011) = 0.5\,\mu P(110) + 0.5\,\mu P(101)$
(002) $\mu P(002) = 0.5\,\mu P(101)$

The normalization condition is

$$P(110) + P(011) + P(200) + P(020) + P(002) + P(101) = 1$$

Solving this set of equations, we have

$$P(002) = \frac{1}{11}, \quad P(101) = \frac{2}{11}, \quad P(011) = \frac{1}{11}$$

$$P(110) = \frac{2}{11}, \quad P(200) = \frac{4}{11}, \quad P(020) = \frac{1}{11}$$

The utilization for Queue 1:

$$\rho_1 = P(110) + P(200) + P(101)$$
$$= \frac{2}{11} + \frac{4}{11} + \frac{2}{11} = \frac{8}{11}$$

7.3 CONVOLUTION ALGORITHM

Before we proceed to evaluate performance measures of a closed queueing network using $P(\bar{n})$, as we did in the case of a single queueing system, let us examine the problem faced with calculating $G_M(N)$, which is part of the expression of $P(\bar{n})$. As pointed out earlier, the state space for a closed queueing network with M queues and N customers is given by

$$|S| = \binom{N + M - 1}{M - 1} = \binom{N + M - 1}{N} \tag{7.11}$$

So for a network with moderate M and N, it is rather computationally expensive to evaluate $G_M(N)$ and often the process will quickly run into the usual combinatorial problem. In the 1970s, Buzen developed an efficient recursive algorithm for computing the normalization constants (Buzen 1973). This somehow reduces the computational burden of evaluating performance measures for a closed queueing network. However, the computation is still substantial, as we shall see later. Here we describe the so-called *convolution algorithm*, which is due to (Buzen 1973), for the case where queues in the network are not state-dependent, that is, we have constant μ_i.

Recall that

$$G_M(N) = \sum_{\bar{n} \in S} \prod_{i=1}^{M} \left(\frac{e_i}{\mu_i} \right)^{n_i} \tag{7.12}$$

If we define a function $g(n,m)$ similar to $G_M(N)$ in definition, we use the lower case n and m to signify that $g_m(n)$ can later be expressed as a recursive function in terms of $g_{m-1}(n)$ and $g_m(n-1)$. By considering the case when the last queue (M) is empty and when it has customers, it can be shown that

$$g_m(n) = g_{m-1}(n) + \left(\frac{e_m}{\mu_m} \right) g_m(n-1) \tag{7.13}$$

This is a convolution-like recursive expression enabling us to compute $g_m(n)$ progressively using the following initial conditions:

$$g_1(n) = (e_1/\mu_1)^n \quad n = 1, 2, \ldots, N \tag{7.14}$$

$$g_m(0) = 1 \quad m = 1, 2, \ldots, M \tag{7.15}$$

Example 7.3

Consider a closed queueing network of three parallel queues, as shown in Figure 7.3. The service times at all three queues are exponentially distributed with mean μ^{-1} and the branching probability $p = 1/2$.

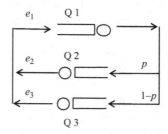

Figure 7.3 A closed network of three parallel queues

Table 7.1 Normalization constants for Figure 7.3 when $e_1 = \mu$

loads (N)	Queues (M)		
	1	**2**	**3**
0	1	1	1
1	$(1)^1 = 1$	$1 + \dfrac{1}{2} \cdot 1 = \dfrac{3}{2}$	$\dfrac{3}{2} + \dfrac{1}{2} \cdot 1 = \dfrac{4}{2}$
2	$(1)^2 = 1$	$1 + \dfrac{1}{2} \cdot \dfrac{3}{2} = \dfrac{7}{4}$	$\dfrac{7}{4} + \dfrac{1}{2} \cdot \dfrac{4}{2} = \dfrac{11}{4}$
3	$(1)^3 = 1$	$1 + \dfrac{1}{2} \cdot \dfrac{7}{4} = \dfrac{15}{8}$	$\dfrac{15}{8} + \dfrac{1}{2} \cdot \dfrac{11}{4} = \dfrac{26}{8}$
4	$(1)^4 = 1$	$1 + \dfrac{1}{2} \cdot \dfrac{15}{8} = \dfrac{31}{16}$	$\dfrac{31}{16} + \dfrac{1}{2} \cdot \dfrac{26}{8} = \dfrac{57}{16}$
5	$(1)^5 = 1$	$1 + \dfrac{1}{2} \cdot \dfrac{31}{16} = \dfrac{63}{32}$	$\dfrac{63}{32} + \dfrac{1}{2} \cdot \dfrac{57}{16} = \dfrac{120}{32}$
6	$(1)^6 = 1$	$1 + \dfrac{1}{2} \cdot \dfrac{63}{32} = \dfrac{127}{64}$	$\dfrac{127}{64} + \dfrac{1}{2} \cdot \dfrac{120}{32} = \dfrac{247}{64}$

(i) By the continuity of flow, we have $e_1 = e_2 + e_3$ and $e_2 = e_3 = (1/2)e_1$, since these e's are arbitrary constants and can be found by fixing one of them to a convenient value. Without loss of generality, we select $e_1 = \mu$ and so we have

$$\frac{e_1}{\mu_1} = 1 \quad \text{and} \quad \frac{e_2}{\mu_2} = \frac{e_3}{\mu_3} = \frac{1}{2}$$

The progressive values of $g_m(n)$ are shown in Table 7.1. The entries in the first row are just $g_m(0)$, one of the initial conditions, and all equal to 1. The entries in the first column are also the initial condition given by $g_1(n) = (e_1/\mu_1)^n$. The other entries are the results of applying recursive Equation (7.13).

The general expression for $g_3(n)$ is given by the following expression. The derivation of it is left as an exercise:

$$g_3(n) = 4 - \left(\frac{1}{2}\right)^n (n + 3)$$

(ii) Let us now select $e_1 = 1/2\mu$ and we have

$$\frac{e_2}{\mu_2} = \frac{e_3}{\mu_3} = \frac{1}{4}.$$

Table 7.2 Normalization constants for Figure 7.3 when $e_1 = \frac{1}{2}\mu$

loads (N)	Queues (M)		
	1	**2**	**3**
0	1.0	1.0	1.0
1	$\left(\dfrac{1}{2}\right)^1 = \dfrac{1}{2}$	$\dfrac{1}{2} + \dfrac{1}{4} \cdot 1 = \dfrac{3}{4}$	$\dfrac{3}{4} + \dfrac{1}{4} \cdot 1 = \dfrac{4}{4}$
2	$\left(\dfrac{1}{2}\right)^2 = \dfrac{1}{4}$	$\dfrac{1}{4} + \dfrac{1}{4} \cdot \dfrac{3}{4} = \dfrac{7}{16}$	$\dfrac{7}{16} + \dfrac{1}{4} \cdot \dfrac{4}{4} = \dfrac{11}{16}$
3	$\left(\dfrac{1}{2}\right)^3 = \dfrac{1}{8}$	$\dfrac{1}{8} + \dfrac{1}{4} \cdot \dfrac{7}{16} = \dfrac{15}{64}$	$\dfrac{15}{64} + \dfrac{1}{4} \cdot \dfrac{11}{16} = \dfrac{26}{64}$
4	$\left(\dfrac{1}{2}\right)^4 = \dfrac{1}{16}$	$\dfrac{1}{16} + \dfrac{1}{4} \cdot \dfrac{15}{64} = \dfrac{31}{256}$	$\dfrac{31}{256} + \dfrac{1}{4} \cdot \dfrac{26}{64} = \dfrac{57}{256}$
5	$\left(\dfrac{1}{2}\right)^5 = \dfrac{1}{32}$	$\dfrac{1}{32} + \dfrac{1}{4} \cdot \dfrac{31}{256} = \dfrac{63}{1024}$	$\dfrac{63}{1024} + \dfrac{1}{4} \cdot \dfrac{57}{256} = \dfrac{120}{1024}$
6	$\left(\dfrac{1}{2}\right)^6 = \dfrac{1}{64}$	$\dfrac{1}{64} + \dfrac{1}{4} \cdot \dfrac{63}{1024} = \dfrac{127}{4096}$	$\dfrac{127}{4096} + \dfrac{1}{4} \cdot \dfrac{120}{1024} = \dfrac{247}{4096}$

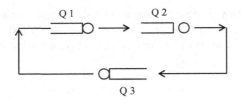

Figure 7.4 A closed serial network

The various normalization constants are shown in Table 7.2.
The general expression for the normalization constants is

$$g_3(n) = \frac{2^{n+2} - n - 3}{4^n}$$

Example 7.4

We next consider a closed queueing network of three serial queues, as shown in Figure 7.4. The service times at all the queues are all exponentially distributed with mean μ^{-1}.

Table 7.3 Normalization constants for Figure 7.4

loads (N)	Queues (M)		
	1	2	3
0	1.0	1.0	1.0
1	$\left(\dfrac{1}{2}\right)^1 = \dfrac{1}{2}$	$\dfrac{1}{2} + \dfrac{1}{2} \cdot 1 = \dfrac{2}{2}$	$\dfrac{2}{2} + \dfrac{1}{2} \cdot 1 = \dfrac{3}{2}$
2	$\left(\dfrac{1}{2}\right)^2 = \dfrac{1}{4}$	$\dfrac{1}{4} + \dfrac{1}{2} \cdot \dfrac{2}{2} = \dfrac{3}{14}$	$\dfrac{3}{4} + \dfrac{1}{2} \cdot \dfrac{3}{2} = \dfrac{6}{4}$
3	$\left(\dfrac{1}{2}\right)^3 = \dfrac{1}{8}$	$\dfrac{1}{8} + \dfrac{1}{2} \cdot \dfrac{3}{4} = \dfrac{4}{8}$	$\dfrac{4}{8} + \dfrac{1}{2} \cdot \dfrac{6}{4} = \dfrac{10}{8}$
4	$\left(\dfrac{1}{2}\right)^4 = \dfrac{1}{16}$	$\dfrac{1}{16} + \dfrac{1}{2} \cdot \dfrac{4}{8} = \dfrac{5}{16}$	$\dfrac{5}{16} + \dfrac{1}{2} \cdot \dfrac{10}{8} = \dfrac{15}{16}$
5	$\left(\dfrac{1}{2}\right)^5 = \dfrac{1}{32}$	$\dfrac{1}{32} + \dfrac{1}{2} \cdot \dfrac{51}{16} = \dfrac{6}{32}$	$\dfrac{9}{32} + \dfrac{1}{2} \cdot \dfrac{15}{16} = \dfrac{21}{32}$
6	$\left(\dfrac{1}{2}\right)^6 = \dfrac{1}{64}$	$\dfrac{1}{64} + \dfrac{1}{2} \cdot \dfrac{6}{32} = \dfrac{7}{64}$	$\dfrac{7}{64} + \dfrac{1}{2} \cdot \dfrac{21}{32} = \dfrac{28}{64}$

In this case, we have $e_1 = e_2 = e_3$. Let us fix the $e_1 = (1/2)\mu$ and we have $(e_i/\mu) = 1/2$. The progressive values of $g_m(n)$ are shown in Table 7.3.

7.4 PERFORMANCE MEASURES

From our early discussions of single-queue systems, we know that once the steady-state probability distribution is obtained we can calculate the queueing parameters that are of interest to us. As pointed out earlier, this approach could lead to computational problems for even moderately sized closed queueing networks. Fortunately, a number of important performance measures can be computed as functions of the various normalization constants $G_M(N)$, which can be evaluated by the recursive convolution algorithm.

In this section, we summarize some performance measures of a closed queueing network in terms of $G_M(N)$ for the case of state-independent queues:

(i) The number of customers at Queue i (Marginal queue length)
By definition, the marginal distribution is

$$P_i(n) = P[N_i = n] = \sum_{\substack{\tilde{n} \in S \\ n_i = n}} P(\tilde{n}) \quad i = 1, 2, \ldots M \tag{7.16}$$

From Chapter 1 Equation (1.8), we see that $P_i(n)$ can also be computed as

$$P_i(n) = P[N_i \geq n] - P[N_i \geq n+1] \tag{7.17}$$

However

$$P[N_i \geq n] = \sum_{\substack{\tilde{n} \in S \\ n_i \geq n}} P(\tilde{n}) = \sum_{\substack{\tilde{n} \in S \\ n_i \geq n}} \frac{1}{g_M(N)} \prod_{j=1}^{M} \left(\frac{e_j}{\mu_j}\right)^{n_j} \tag{7.18}$$

Since Queue i has more than n customers, we can factor out $(e_i/\mu_i)^n$ from the last product term, and what remains is simply the normalization constant with $N - n$ customers. Hence

$$P[N_i \geq n] = \left(\frac{e_i}{\mu_i}\right)^n \frac{g_M(N-n)}{g_M(N)} \tag{7.19}$$

Substituting back into Equation (7.17), we have

$$P_i(n) = \left(\frac{e_i}{\mu_i}\right)^n \frac{1}{g_M(N)} \left[g_M(N-n) - \left(\frac{e_i}{\mu_i}\right) g_M(N-n-1)\right] \tag{7.20}$$

(ii) The mean number of customers at Queue i
The mean number of customers at Queue i is by definition

$$E[N_i] = \sum_{k=1}^{N} k P[N_i = k] \tag{7.21}$$

But from Example 1.8 in Chapter 1, we know that the expected value of a discrete random variable can alternatively be computed as

$$E[N_i] = \sum_{k=1}^{N} P[N_i \geq k]$$

$$= \sum_{k=1}^{N} \left(\frac{e_i}{\mu_i}\right)^k \frac{g_M(N-k)}{g_M(N)} \tag{7.22}$$

(iii) Marginal utilization of node i
For a single-server queueing network, the marginal utilization of node i is

$$\rho_i = 1 - P[N_i = 0] = P[N_i \geq 1]$$

$$= \left(\frac{e_i}{\mu_i}\right) \frac{g_M(N-1)}{g_M(N)} \tag{7.23}$$

(iv) Marginal throughput of node i

For a single-server and state-independent network, the throughput is by definition

$$\Psi_i(N) = \sum_{k=1}^{N} \mu_i P[N_i = k]$$

$$= \mu_i P[N_i \geq 1]$$

$$= \mu_i \left[\left(\frac{e_i}{\mu_i} \right) \frac{g_M(N-1)}{g_M(N)} \right]$$

$$= e_i \frac{g_M(N-1)}{g_M(N)} \tag{7.24}$$

(v) The mean waiting time at Queue i

$$T_i = \frac{E(N_i)}{\Psi_i} = \frac{\sum_{k=1}^{N} (e_i/\mu_i)^k g_M(N-k)}{e_i g_M(N-1)} \tag{7.25}$$

Example 7.5

Let us now calculate some performance measures of Queue 1 for the network shown in Example 7.3:

(i) For the case of $e_1/\mu_1 = 1$ and $e_2/\mu_2 = e_3/\mu_3 = 1/2$, let us assume that $N = 5$. We have from Section 7.4:

$$P[N_1 = 2] = \left(\frac{e_i}{\mu_i} \right)^n \frac{1}{g_M(N)} \left[g_M(N-n) - \left(\frac{e_i}{\mu_i} \right) g_M(N-n-1) \right]$$

$$= (1)^2 \frac{32}{120} \left[\frac{26}{8} - (1)\frac{11}{4} \right] = \frac{2}{15}$$

$$E[N_1] = \sum_{k=1}^{N} \left(\frac{e_i}{\mu_i} \right)^k \frac{g_M(N-k)}{g_M(N)}$$

$$= \frac{32}{120} \left[\frac{57}{16} + \frac{26}{8} + \frac{11}{4} + \frac{4}{2} + 1 \right] = \frac{67}{20}$$

$$\Psi_1(5) = e_i \frac{g_M(N-k)}{g_M(N)}$$

$$= \mu \left[\frac{57}{16} \times \frac{32}{120} \right] = \frac{19}{20} \mu$$

(ii) In the case of $e_1/\mu_1 = 1/2$ and $e_2/\mu_2 = e_3/\mu_3 = 1/4$ and $N = 5$:

$$P[N_1 = 2] = \left(\frac{1}{2}\right)^2 \frac{1024}{120} \left[\frac{26}{64} - \left(\frac{1}{2}\right) \times \frac{11}{16}\right] = \frac{2}{15}$$

$$E(N_1) = \frac{1024}{120} \left[\left(\frac{1}{2}\right)\frac{57}{256} + \left(\frac{1}{2}\right)^2\frac{26}{64} + \left(\frac{1}{2}\right)^3\frac{11}{16} + \left(\frac{1}{2}\right)^4\frac{4}{4} + \left(\frac{1}{2}\right)^5 1\right]$$

$$= \frac{67}{20}$$

$$\Psi_1(5) = \mu\left[\frac{57}{256} \times \frac{1024}{120}\right] = \frac{19}{20}\mu$$

7.5   MEAN VALUE ANALYSIS

From the preceding examples, we see that substantial computation is involved in evaluating the performance measures of a closed queueing network. Though the convolution algorithm is meant to be implemented as a computer program, the rapidly growing values of the progressive $g_m(n)$ can cause overflow/underflow problems in the computing process.

In this section we present another technique called Mean Value Analysis (MVA), which tackles the performance issue without the explicit evaluation of normalization constants. This technique is due to Reiser and Lavenberg (Reiser and Lavenberg 1980; Reiser 1982).

For simplicity, we shall look at the case where the service rates μ_i are state independent. Again, focusing on the time a customer spends at a particular queue, say the ith queue, we see that his/her queueing time is equal to his/her service time plus that of those customers ahead of him/her. That is

$$T_i = \mu_i^{-1} + \mu_i^{-1} \times (\text{average number of customers upon arrival}) \quad (7.26)$$

Note that the residual service time of the customer in service does not come into the equation because of the exponential service times. It has been shown (Lavenberg and Reiser 1980) that for a closed queueing network that has a product form solution, the number of customers in the queue seen by an arriving customer at node i has the same distribution as that seen by a random outside observer with one customer less. Therefore, the preceding equation can be written as

$$T_i(n) = \mu_i^{-1}[1 + N_i(n-1)] \quad (7.27)$$

where $N_i(n)$ is the average number of customers in the ith queue when there are n customers in the network.

Now, let us apply Little's theorem to the network as a whole. We know that the total number of customers in the network is n and the mean time a customer spends in the network is simply the sum of the time he/she spends at each of the individual nodes:

$$n = \Psi(n) \sum_{i=1}^{M} \overline{T_i}(n) \tag{7.28}$$

where $\Psi(n)$ is the system throughput and $\overline{T_i}(n)$ is the average total time a customer spends at node i. Students should not confuse $\overline{T_i}(n)$ with T_i because a customer may visit a node more than once. T_i is the time a customer spends in a node per visit, whereas $\overline{T_i}(n)$ is the total time he/she spends at node i.

To obtain an expression for $\overline{T_i}(n)$, recall that the flow equation of a closed queueing network is given by

$$\lambda_i = \sum_{j=1}^{M} \lambda_j p_{ji} \quad i = 1, 2, \ldots, M \tag{7.29}$$

If we fix $e_1 = 1$, the solution e_i for the above equation can be interpreted as the mean number of visits to node i by a customer (Denning and Buxen 1978). Thus we have

$$\overline{T_i}(n) = e_i T_i(n) \tag{7.30}$$

Substituting into expression (7.28), we arrive at

$$n = \Psi(n) \sum_{i=1}^{M} e_i T_i(n) \tag{7.31}$$

The suggested set of solution ($e_1 = 1$) also implies that

$$\Psi_1(n) = \Psi(n) \tag{7.32}$$

$$\Psi_j(n) = \Psi_1(n) e_j \quad j = 2, 3, \ldots, M \tag{7.33}$$

Collecting Equations (7.27), (7.32) and (7.33), we have the set of recursive algorithms for the network:

$$N_i(0) = 0 \quad i = 1,2,3, \ldots, M$$

$$T_i(n) = \mu_i^{-1}[1 + N_i(n - 1)$$

$$\Psi_1(n) = n \times \left[\sum_{i=1}^{M} e_i T_i(n) \right]^{-1}$$

$$\Psi_j = \Psi_i(n)e_i \quad j = 2,3, \ldots M$$

$$N_i(n) = \Psi_i(n)T_i(n)$$

Example 7.6: Mean Value Analysis

We shall use Example 7.4 and calculate some of the performance parameters. Since all the queues are connected in series, we have $\Psi_i(n) = \Psi(n)$ and the set of MVA algorithm reduced to

$$N_i(0) = 0 \quad i = 1,2,3, \ldots, M$$

$$T_i(n) = \mu_i^{-1}[1 + N_i(n - 1)]$$

$$\Psi(n) = n \times \left[\sum_{i=1}^{M} T_i(n) \right]^{-1}$$

$$N_i(n) = \Psi(n)T_i(n)$$

We are given $\mu_i^{-1} = \mu^{-1}$, therefore we have

(i)   First iteration, i.e. 1 customer in the network:

$$\bar{n}_i(0) = 0 \quad i = 1,2,3$$

$$T_i(1) = \mu^{-1}[1 + \bar{n}_i(0)] = \mu^{-1}$$

$$\Psi(1) = 1 \times \left[\sum_{i=1}^{3} T_i(1) \right]^{-1} = \frac{\mu}{3}$$

$$N_i(1) = \Psi(1) \times \mu^{-1} = 1/3$$

(ii) Second iteration: i.e. 2 customers in the network:

$$T_i(2) = \mu^{-1}[1 + (1/3)] = 4/3 \, \mu$$

$$\Psi(2) = 2 \times \left[\sum_{i=1}^{3} T_i(2) \right]^{-1} = \frac{\mu}{2}$$

$$N_i(2) = \Psi(2) \times (4/3 \, \mu) = 2/3$$

(iii) Third iteration: i.e. 3 customers in the network:

$$T_i(3) = \mu^{-1}[1 + (2/3)] = 5/3 \, \mu$$

$$\Psi(3) = 3 \times \left[\sum_{i=1}^{3} T_i(3) \right]^{-1} = \frac{3}{5}\mu$$

$$N_i(3) = \Psi(3) \times (5/3 \, \mu) = 1$$

7.6   APPLICATIONS OF CLOSED QUEUEING NETWORKS

Closed queueing networks are well suited for modelling batch processing computer systems with a fixed level of multi-programming. A computer system can be modelled as a network of interconnected resources (CPU and I/O devices) and a collection of customers (computer jobs), as shown in Figure 7.5. This queueing model is known as the Central Server System in classical queueing theory.

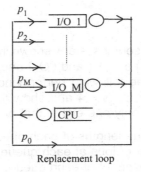

Replacement loop

Figure 7.5   A central server queueing system

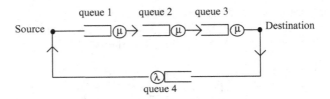

Figure 7.6   Queueing model for a virtual circuit

When a job returns immediately back to CPU without going through the I/O devices, it is considered as having completed its task and is replaced by another job, as shown by the replacement loop.

Another application of closed queueing network models is the quantitative analysis of a sliding window flow control mechanism as applied to a virtual circuit of a data network. Consider the case where end-to-end window flow control is applied to VC1 of the network shown in Figure 6.6 of Chapter 6. By virtue of the window size N, there can be at most N packets traversing along VC1 at any one time. If we further assume that each packet is only acknowledged individually at the destination and these end-to-end acknowledgements are accorded the highest priority and will come back through the network with no or little delays, then the VC1 can then be modelled as a closed queueing network, as shown in Figure 7.6.

An artificial Queue 4 is added between the source and destination to reflect the fact that if there are N packets along the VC (from node 1 to node 3), Queue 4 is empty and no additional packets can be admitted into VC. However, when a packet arrives at the destination, it appears immediately at the 4th queue and hence one additional packet is added to the VC. For the detailed queueing modelling and analysis of sliding window flow control, students are directed to Chapter 9 on the application of closed queueing networks.

Problems

1. Consider the 'central server system' shown in Figure 7.5. If there are only two I/O queues, i.e. I/O 1 and I/O 2 and $p_0 = 0.125$, $p_1 = 0.375$, $p_2 = 0.5$, calculate the following performance measures using the convolution algorithm for $N = 4$ and show that the service rates of both I/O 1 and 2 are only one-eighth that of the CPU queue:

   (i)   the marginal queue lengths of each queue in the network;
   (ii) the mean number of jobs at each queue;
   (iii) the marginal throughput of each queue;
   (iv) the mean waiting time at each queue.

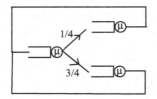

Figure 7.7 A closed network of three queues

2. Calculate the performance measure (ii) to (iv) for the previous problem using the Mean Value Analysis method.
3. Refer to Example 7.3. Show that the general expressions of $g_3(n)$ are as follows for part (i) and part (ii) of the problem:

$$g_3(n) = 4 - \left(\frac{1}{2}\right)^n (n+3)$$

$$g_3(n) = \frac{2^{N+2} - N - 3}{4^N}$$

4. Consider the following closed queueing network (Figure 7.7) with only two customers; find $P(k_1, k_2, k_3)$ explicitly in terms of μ.

8

Markov-Modulated Arrival Process

In all the queueing systems (or networks) that we have discussed so far, we have always adopted the rather simple arrival model – Poisson or General Independent process that specifically assumes arrivals occur independently of one another and of the service process. This simple assumption is not always adequate in modelling real-life data traffic, which exhibits correlations in their arrivals, especially the new services found in Asynchronous Transfer Mode (ATM) or broadband networks.

ATM is a transport mechanism recommended for the Broadband Integrated Services Digital Network that is supposedly capable of handling a wide mixture of traffic sources; ranging from the traditional computer data, to packetized voice and motion video. In this multi-media traffic, the arrival instants are in some sense correlated and exhibit a diverse mixture of traffic characteristics with different correlations and burstiness. As queueing system performances are highly sensitive to these correlations and bursty characteristics in the arrival process, we need more realistic models to represent the arrival process.

In this chapter, we first examine three traffic models, namely:

- Markov-modulated Poisson Process (MMPP)
- Markov-modulated Bernoulli process (MMBP), and
- Markov-modulated Fluid Flow.

These traffic models are widely used as source models in performance evaluation of multi-media traffic due to their ability to capture the strong correlation in traffic intensity in close proximity in time, and yet being analytically tractable.

Later in the chapter, we will look at a new paradigm of queueing analysis called deterministic queueing, or so-called *network calculus* by Jean-Yves Boudec. This new paradigm of queueing analysis is able to put deterministic bounds on network performance measures.

8.1   MARKOV-MODULATED POISSON PROCESS (MMPP)

MMPP is a term introduced by Neuts (Neuts 1989) for a special class of versatile point processes whose Poisson arrivals are modulated by a Markov process. This class of models was earlier studied by Yechiali and Naor who named them M/M/1 queues with heterogeneous arrivals and services.

MMPP has been used to model various multi-media sources which have time-varying arrival rates and correlations between inter-arrival time, such as packetized voice and video, as well as their superposed traffic. It still remains tractable analytically and produces fairly good results.

8.1.1   Definition and Model

In the MMPP model, arrivals are generated by a source whose stochastic behaviour is governed by an *m-state* irreducible continuous-time Markov process, which is independent of the arrival process.

While the underlying modulating Markov process is spending an exponentially distributed time in state i ($i = 1, 2, \ldots, m$), the MMPP is said to be in state $J(t) = i$ and arrivals are generated according to a Poisson process with rate λ_i, as shown in Figure 8.1.

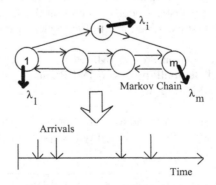

Figure 8.1   Markov-modulated Poisson process

The Markov-modulated Poisson process (MMPP) is a doubly stochastic Poisson process whose rate varies according to a continuous time function. In this case, the time function is the underlying Markov chain. MMPP is fully characterized by the following parameters:

(i) The transition-rate matrix (also known as the infinitesimal generator $\tilde{Q}$) of the underlying modulating Markov process:

$$\tilde{Q} = \begin{bmatrix} -q_1 & q_{12} & \cdots & q_{1m} \\ q_{21} & -q_2 & \cdots & q_{2m} \\ \cdots & & & \cdots \\ q_{m1} & & & -q_m \end{bmatrix} \quad (8.1)$$

where

$$q_i = \sum_{\substack{j=1 \\ j \neq i}}^{m} q_{ij}$$

We assume here that the transition-rate matrix $\tilde{Q}$ is homogeneous, i.e. $\tilde{Q}$ does not vary with time. The steady-state vector:

$$\tilde{\pi} = [\pi_1, \pi_2, \ldots, \pi_m]$$

of the modulating Markov chain is then given by

$$\tilde{\pi}\tilde{Q} = 0 \quad and \quad \pi_1 + \pi_2 + \ldots + \pi_m = 1$$

(ii) The Poisson arrival rate at each state $\lambda_1, \lambda_2, \ldots, \lambda_m$. We define a diagonal matrix $\tilde{\Lambda}$ and a vector $\tilde{\lambda}$ as

$$\tilde{\Lambda} = diag(\lambda_1, \lambda_2, \ldots \lambda_m)$$

$$= \begin{bmatrix} \lambda_1 & 0 & \cdots & 0 \\ 0 & \lambda_2 & \cdots & 0 \\ \cdots & & & 0 \\ 0 & 0 & \cdots & \lambda_m \end{bmatrix} \quad (8.2)$$

$$\tilde{\lambda} = (\lambda_1, \lambda_2, \ldots \lambda_m)^T \quad (8.3)$$

(iii) The initial state (initial probability vector) of the MMPP, i.e.:

$$\varphi_i = P[J(t=0)=i] \quad \& \quad \sum_i \varphi_i = 1.$$

Depending on the initial vector chosen, we have

(a) an interval-stationary MMPP that starts at an 'arbitrary' arrival epoch if the initial vector is chosen as

$$\tilde{\varphi} = \frac{1}{\pi_1 \lambda_1 + \pi_2 \lambda_2 + \ldots + \pi_m \lambda_m} \tilde{\pi} \tilde{\Lambda} \tag{8.4}$$

(b) an environment stationary MMPP whose initial probability vector is chosen as $\tilde{\pi}$, which is the stationary vector of $\tilde{Q}$. Now the origin of time is not an arrival epoch, but is chosen so that the environmental Markov process is stationary.

Having described the model, let us examine the distribution of the inter-arrival time of an MMPP. If we denote X_k as the time between $(k-1)$ arrival and k arrival, and J_k the state of the underlying Markov process at the time of the kth arrival, then the sequence $\{(J_k, X_k), k \geq 0\}$ is a Markov renewal sequence with a transition probability distribution matrix given by

$$\tilde{F}(x) = \int_0^x e^{[(\tilde{Q}-\tilde{\Lambda})\varsigma]} d\varsigma \tilde{\Lambda}$$

$$= [-e^{(\tilde{Q}-\tilde{\Lambda})\varsigma} (\tilde{\Lambda}-\tilde{Q})^{-1}]_0^x \tilde{\Lambda}$$

$$= \{\tilde{I} - e^{(\tilde{Q}-\tilde{\Lambda})x}\}(\tilde{\Lambda}-\tilde{Q})^{-1} \tilde{\Lambda} \tag{8.5}$$

The elements $F_{ij}(x)$ of the matrix are the conditional probabilities:

$$F_{ij}(x) = P\{J_k = j, X_k \leq x | J_{k-1} = i\} \tag{8.6}$$

The transition probability density matrix is then given by differentiating $\tilde{F}(x)$ with respect to x:

$$\tilde{f}(x) = \frac{d}{dx} \tilde{F}(x) = \{-e^{-(\tilde{Q}-\tilde{\Lambda})x}(\tilde{Q}-\tilde{\Lambda})\}(\tilde{\Lambda}-\tilde{Q})^{-1} \tilde{\Lambda}$$

$$= e^{(\tilde{Q}-\tilde{\Lambda})x} \tilde{\Lambda} \tag{8.7}$$

Taking the Laplace transform of $\tilde{f}(x)$, we have

$$L[\tilde{f}(x)] = \int_0^\infty e^{(\tilde{Q}-\tilde{\Lambda})x} e^{-sx} dx \tilde{\Lambda}$$

$$= [-(s\tilde{I} - \tilde{Q} + \tilde{\Lambda})^{-1} e^{-(s\tilde{I}-\tilde{Q}+\tilde{\Lambda})x}]_0^\infty \tilde{\Lambda}$$

$$= (s\tilde{I} - \tilde{Q} + \tilde{\Lambda})^{-1} \tilde{\Lambda} \tag{8.8}$$

Now let us consider $N(t)$, the number of arrivals in $(0, t)$. If $J(t)$ is the state of the Markov process at time t, define a matrix $\tilde{P}(n, t)$ whose (i, j) entry is

$$P_{ij}(n, t) = P[N(t) = n, J(t) = j | N(t = 0) = 0, J(t = 0) = i] \tag{8.9}$$

Then the matrices satisfy the Chapman–Kolmogorov equation and the matrix generating function $\tilde{P}*(z, t)$ is

$$\tilde{P}*(z, t) = \sum_{n=0}^\infty \tilde{P}(n, t) z^n$$

$$= \exp\{(\tilde{Q} - (1-z)\tilde{\Lambda})t\} \tag{8.10}$$

with

$$\tilde{P}*(z, 0) = \tilde{I} \tag{8.11}$$

then the probability generating function of the number of arrivals is

$$g(n, t) = \tilde{\pi}\{\exp(\tilde{Q} - (1-z)\tilde{\Lambda})t\}\tilde{e} \tag{8.12}$$

where $\tilde{\pi}$ is the steady-state probability vector of the Markov chain and $\tilde{e} = [1, 1, \dots 1]^T$.

Example 8.1

A two-state MMPP has only two states in its underlying modulating Markov chain, characterized by

$$\tilde{Q} = \begin{bmatrix} -r_1 & r_1 \\ r_2 & -r_2 \end{bmatrix} \quad \text{and} \quad \tilde{\Lambda} = \begin{bmatrix} \lambda_1 & 0 \\ 0 & \lambda_2 \end{bmatrix}$$

The stationary distribution is given by

$$\tilde{\pi} = \begin{bmatrix} \dfrac{r_2}{r_1 + r_2}, & \dfrac{r_1}{r_1 + r_2} \end{bmatrix}.$$

From Equation (8.8), we obtain

$$L[\tilde{f}(x)] = \left\{ \begin{pmatrix} s & 0 \\ 0 & s \end{pmatrix} - \begin{pmatrix} -r_1 & r_1 \\ r_2 & -r_2 \end{pmatrix} + \begin{pmatrix} \lambda_1 & 0 \\ 0 & \lambda_2 \end{pmatrix} \right\}^{-1} \begin{pmatrix} \lambda_1 & 0 \\ 0 & \lambda_2 \end{pmatrix}$$

$$= \frac{1}{\det A} \begin{pmatrix} s + r_2 + \lambda_2 & r_1 \\ r_2 & s + r_1 + \lambda_1 \end{pmatrix} \begin{pmatrix} \lambda_1 & 0 \\ 0 & \lambda_2 \end{pmatrix}$$

$$= \frac{1}{\det A} \begin{pmatrix} (s + r_2 + \lambda_2)\lambda_1 & \lambda_2 r_1 \\ \lambda_1 r_2 & (s + r_1 + \lambda_1)\lambda_2 \end{pmatrix}$$

where

$$\det A = (s + r_1 + \lambda_1)(s + r_2 + \lambda_2) - r_1 r_2.$$

Without loss of generality, let us assume that this MMPP starts at an arrival epoch. The stationary distribution of the Markov chain embedded at the arrival instants is

$$\tilde{\varphi} = \frac{\tilde{\pi}\tilde{\Lambda}}{\tilde{\pi}\tilde{\lambda}} = \frac{1}{r_2 \lambda_1 + r_1 \lambda_2}[r_2 \lambda_1 \quad r_1 \lambda_2]$$

Hence, the Laplace transform of the unconditional inter-arrival time is then

$$L[X] = \tilde{\varphi} \left\{ \frac{1}{\det A} \begin{pmatrix} (s + r_2 + \lambda_2)\lambda_1 & \lambda_2 r_1 \\ \lambda_1 r_2 & (s + r_1 + \lambda_1)\lambda_2 \end{pmatrix} \right\} \begin{pmatrix} 1 \\ 1 \end{pmatrix}$$

$$= \frac{s(r_1\lambda_2^2 + r_2\lambda_1^2) + (r_1\lambda_2 + r_2\lambda_1)(r_1\lambda_2 + \lambda_1\lambda_2 + r_2\lambda_1)}{(r_1\lambda_2 + r_2\lambda_1)\{s^2 + (r_1 + r_2 + \lambda_1 + \lambda_2)s + (r_1\lambda_2 + \lambda_1\lambda_2 + r_2\lambda_1)\}}$$

The interrupted Poisson process (IPP) is a special case of the two-state MMPP, characterized by the following two matrices:

$$\tilde{Q}_{IPP} = \begin{bmatrix} -r_1 & r_1 \\ r_2 & -r_2 \end{bmatrix} \quad \text{and} \quad \tilde{\Lambda}_{IPP} = \begin{bmatrix} \lambda & 0 \\ 0 & 0 \end{bmatrix}$$

By setting $\lambda_1 = \lambda$ and $\lambda_2 = 0$ in the previous expression, we have the Laplace transform of the inter-arrival times of an IPP as

$$L[X_{IPP}] = \frac{(s + r_2)\lambda}{s^2 + (r_1 + r_2 + \lambda)s + r_2\lambda}$$

$$= (1 - \beta)\frac{\alpha_1}{s + \alpha_1} + \beta\frac{\alpha_2}{s + \alpha_2}$$

where

$$\alpha_1 = \frac{1}{2}\left[(r_1 + r_2 + \lambda) - \sqrt{(r_1 + r_2 + \lambda)^2 - 4r_2\lambda}\right]$$

$$\alpha_2 = \frac{1}{2}\left[(r_1 + r_2 + \lambda) + \sqrt{(r_1 + r_2 + \lambda)^2 - 4r_2\lambda}\right]$$

$$\beta = \frac{\lambda - \alpha_1}{\alpha_2 - \alpha_1}$$

We note the expression for $L[X_{IPP}]$ is just the Laplace transform of a hyper-exponential distribution with parameters (α_1, α_2) and branching probability β. Therefore, the interrupted Poisson process is stochastically equivalent to a hyper-exponential process.

8.1.2 Superposition of MMPPs

The superposition of $n \geq 2$ independent Markov-modulated Poisson processes with parameters $\tilde{Q}\&\tilde{\Lambda}_i (i = 1,2, \ldots, n)$ is again an MMPP with parameters $\tilde{Q}\&\tilde{\Lambda}$ given by

$$\tilde{Q} = \tilde{Q}_1 \oplus \tilde{Q}_2 \oplus \ldots \oplus \tilde{Q}_n \tag{8.13}$$

$$\tilde{\Lambda} = \tilde{\Lambda}_1 \oplus \tilde{\Lambda}_2 \oplus \ldots \oplus \tilde{\Lambda}_n \tag{8.14}$$

where $\oplus$ denotes the Kronecter sum and is defined as

$$\tilde{A} \oplus \tilde{B} = (\tilde{A} \otimes \tilde{I}_B) + (\tilde{I}_A \otimes \tilde{B}) \tag{8.15}$$

and

$$\tilde{C} \otimes \tilde{D} = \begin{bmatrix} c_{11}\tilde{D} & c_{12}\tilde{D} & \cdots & c_{1m}\tilde{D} \\ c_{21}\tilde{D} & c_{22}\tilde{D} & \cdots & c_{2m}\tilde{D} \\ \cdots & \cdots & & \\ c_{n1}\tilde{D} & c_{n2}\tilde{D} & \cdots & c_{nm}\tilde{D} \end{bmatrix} \tag{8.16}$$

$\tilde{I}_A \& \tilde{I}_B$ are identity matrices with the same dimension as A and B, respectively. Note that $\tilde{Q}\&\tilde{\Lambda}$ are $k x k$ matrices and $k = \prod_{i=1}^{n} n_i$.

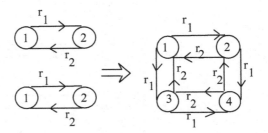

Figure 8.2 Superposition of MMPPs

Example 8.2

Consider the superposition of two identical two-state MMPPs with parameters r_1 and r_2, as shown in Figure 8.2. The Poisson arrival rates in these two states are λ_1 and λ_2, respectively.

Using expressions (8.13) and (8.14), the transition rate matrix $\tilde{Q}$ and arrival rate matrix $\tilde{\Lambda}$ of the combined arrival process is given by

$$\tilde{Q} = \tilde{Q}_1 \oplus \tilde{Q}_2$$

and

$$\tilde{\Lambda} = \tilde{\Lambda}_1 \oplus \tilde{\Lambda}_2$$

Therefore:

$$\tilde{Q} = \begin{bmatrix} -r_1 & r_1 \\ r_2 & -r_2 \end{bmatrix} \otimes \begin{bmatrix} 1 & 0 \\ 0 & 1 \end{bmatrix} + \begin{bmatrix} 1 & 0 \\ 0 & 1 \end{bmatrix} \otimes \begin{bmatrix} -r_1 & r_1 \\ r_2 & -r_2 \end{bmatrix}$$

$$= \begin{bmatrix} -2r_1 & r_1 & r_1 & 0 \\ r_2 & -r_1-r_2 & 0 & r_1 \\ r_2 & 0 & -r_1-r_2 & r_1 \\ 0 & r_2 & r_2 & r_2 \end{bmatrix}$$

$$\tilde{\Lambda} = \begin{bmatrix} \lambda_1 & 0 \\ 0 & \lambda_2 \end{bmatrix} \otimes \begin{bmatrix} 1 & 0 \\ 0 & 1 \end{bmatrix} + \begin{bmatrix} 1 & 0 \\ 0 & 1 \end{bmatrix} \otimes \begin{bmatrix} \lambda_1 & 0 \\ 0 & \lambda_2 \end{bmatrix}$$

$$= \begin{bmatrix} 2r_1 & 0 & 0 & 0 \\ 0 & \lambda_1+\lambda_2 & 0 & r_1 \\ 0 & 0 & \lambda_1+\lambda_2 & r_1 \\ 0 & 0 & 0 & 2\lambda_2 \end{bmatrix}$$

The resultant Markov chain is also shown in Figure 8.2 to have a comparison.

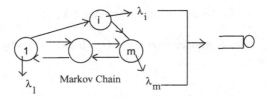

Figure 8.3   MMPP/G/1

8.1.3   MMPP/G/1

This is a queueing system where customers arrive according to an MMPP and are served by a single server with a general service time distribution $x(t)$. The waiting queue is infinite and customers are served according to their order of arrival, as shown in Figure 8.3.

The MMPP is characterized by the matrices $\tilde{Q}\,\&\,\tilde{\Lambda}$, and the service time distribution has a mean $1/\mu$ and Laplace transform $X(s)$. The utilization of the system is

$$\rho = \tilde{\pi}\tilde{\lambda}/\mu = \frac{\pi_1\lambda_1 + \pi_2\lambda_2 + \ldots + \pi_m\lambda_m}{\mu}$$

Consider the successive epochs of departure $\{\tau_k : k \geq 0\}$. Firstly, let us define $N_k\,\&\,J_k$ to be the number of customers in the system and the state of the MMPP at τ_k respectively, then the sequence $\{(N_k, J_k, \tau_{k+1} - \tau_k) : k \geq 0\}$ forms a semi-Markov sequence with the transition probability matrix:

$$\tilde{Q} = \begin{bmatrix} \tilde{B}_0 & \tilde{B}_1 & \tilde{B}_2 & \cdots \\ \tilde{A}_0 & \tilde{A}_1 & \tilde{A}_2 & \cdots \\ 0 & \tilde{A}_0 & \tilde{A}_1 & \cdots \\ \cdots & & & \cdots \end{bmatrix} \tag{8.17}$$

where

$$\tilde{A}_n(\varsigma) = \int_0^\varsigma P(n, t)d(x(t)) \quad \text{and} \quad \tilde{B}_n = (\tilde{\Lambda} - \tilde{Q})^{-1}\tilde{\Lambda}\tilde{A}_n(\infty) \tag{8.18}$$

If we define $w_i(t)$ to be the joint probability that the MMPP is in phase i and that a customer who arrives at that time would wait less than or equal to t before receiving service, then the virtual delay distribution matrix, $\tilde{W}(t) = (w_i(t))$ is given by the following expression in Laplace transform:

$$\tilde{W}(s) = s(1-\rho)\tilde{g}(s\tilde{I} + \tilde{Q} - \tilde{\Lambda} + \tilde{\Lambda}X(s))^{-1} \qquad (8.19)$$

The vector $\tilde{g}$ is the steady-state vector of $\tilde{G}$, which is given by the following implicit equation:

$$\tilde{G} = \int_0^\infty e^{(\tilde{Q}-\tilde{\Lambda}+\tilde{\Lambda}\tilde{G})x} dX(x) \qquad (8.20)$$

8.1.4   Applications of MMPP

The two-state MMPP can be used to model a single packetized voice source, which exhibits two alternate phases, an active or talk spurt (On) period and a silent (Off) period. The sojourn time in each of these two phases is approximated by an exponentially distributed random variable with means $1/\alpha$ and $1/\beta$, respectively. Voice packets are generated during the active period and no packets are generated during the silent period, as shown in Figure 8.4. To model the packets' arrival process, we approximate it by a Poisson arrival process with mean λ, hence the arrival process represents an interrupted Poisson process (IPP), which is a special case of the two-state MMPP.

A *m-state* MMPP, as shown in Figure 8.5, can be used to describe the super-position process of *m* voice sources, each of which is modelled as an On-Off source as before. The state of the Markov chain represents the number of active sources.

It can also be extended to model a video source that switches between two or more distinct modes of operation with different correlations and burstiness coefficients.

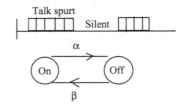

Figure 8.4   Interrupted Poisson Process model for a voice source

Figure 8.5   An m-state MMPP model for voice sources

8.2 MARKOV-MODULATED BERNOULLI PROCESS

The Markov-modulated Bernoulli Process (MMBP) is the discrete-time counter-part of Markov-modulated Poisson Process (MMPP). Time in MMBPs is discretized into fixed-length slots and the process spends a geometric duration of time slots in each state. The probability that a slot containing an arrival is a Bernoulli process, with a parameter that varies according to an m-state Markov process, is independent of the arrival process.

8.2.1 Source Model and Definition

To serve as a teaching example, we consider only the two-state MMBP, as shown in Figure 8.6, to simplify the analysis. In this model, the arrival process has two distinct phases (or states); when the arrival process is in H-state in time slot k, it generates an arrival with probability γ and may remain in this state in the next time slot $(k + 1)$ with probability α. Similarly, when the arrival process is in L-state in time slot k, it generates an arrival with probability ζ and may remain in L-state in the next time slot $(k + 1)$ with probability β. The width of the time slots are taken to be the average service time in subsequent analysis.

From the Markov theory, we know that the steady-state probability distribution $\tilde{\pi} = [\pi_H, \pi_L]$ for this two-state chain is given by $\tilde{\pi}\tilde{P} = \tilde{\pi}$, where $\tilde{P}$ is the state transition matrix:

$$\tilde{P} = \begin{pmatrix} \alpha & 1-\alpha \\ 1-\beta & \beta \end{pmatrix} \tag{8.21}$$

Solving the equation $\tilde{\pi}\tilde{P} = \tilde{\pi}$ together with $\pi_H + \pi_L = 1$, we have

$$\pi_H = \frac{1-\beta}{2-\alpha-\beta} \quad \text{and} \quad \pi_L = \frac{1-\alpha}{2-\alpha-\beta} \tag{8.22}$$

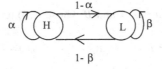

Figure 8.6 A two-state MMBP

8.2.2   Superposition of N Identical MMBPs

Having presented the model of a single MMBP, we are now in a position to examine the stochastic behaviour of superposing (multiplexing) a group of N identical MMBPs. Firstly, let us define the following random variables:

$\alpha(k + 1)$ The number of traffic sources in H-state in slot $(k + 1)$
$a(k)$ The number of traffic sources in H-state in slot k
$b(k)$ The total number of cells generated by these N source in slot k

With a bit of thought, it can be shown that these random variables are related as follows:

$$a(k+1) = \sum_{j=1}^{a(k)} c_j + \sum_{j=1}^{N-a(k)} d_j \qquad (8.23)$$

$$b(k) = \sum_{j=1}^{a(k)} s_j + \sum_{j=1}^{N-a(k)} t_j \qquad (8.24)$$

where c_j, d_j, s_j and t_j are random variables with the following definitions and probability generating functions (PDFs):

$$c_j = \begin{cases} 1 & \text{with probability } \alpha \\ 0 & \text{with probability } (1-\alpha) \end{cases} \qquad \text{PDF is } F_H(z) = (1-\alpha) + \alpha z$$

$$d_j = \begin{cases} 1 & \text{with probability } (1-\beta) \\ 0 & \text{with probability } \beta \end{cases} \qquad \text{PDF is } F_L(z) = \beta + (1-\beta)z$$

$$s_j = \begin{cases} 1 & \text{with probability } \gamma \\ 0 & \text{with probability } (1-\gamma) \end{cases} \qquad \text{PDF is } G_H(z) = (1-\gamma) + \gamma z$$

$$t_j = \begin{cases} 1 & \text{with probability } \varsigma \\ 0 & \text{with probability } (1-\varsigma) \end{cases} \qquad \text{PDF is } G_L(z) = (1-\varsigma) + \varsigma z$$

Next, we define two matrices $\tilde{A}$ and $\tilde{B}$ whose (i, j) terms are

$$A(i, j) = P[a(k+1) = j \,|\, a(k) = i] \qquad (8.25)$$

$$B(i, j) = P[b(k) = j \,|\, a(k) = i] \qquad (8.26)$$

Matrix $\tilde{A}$ gives the correlation between the number of H-state MMBP sources in adjacent time slots, and matrix $\tilde{B}$ shows the relationship between the number of arrivals and the number of H-state MMBP sources in a particular time slot. Both matrices are required in subsequent analysis.

To facilitate the build-up of these two matrices, we see that the z-transform of the ith row of these two matrices are given respectively as

$$A_i(z) = \sum_{j=0}^{\infty} P[a(k+1) = j | a(k) = i] z^n$$
$$= F_H^i(z) \cdot F_L^{N-i}(z) \tag{8.27}$$

$$B_i(z) = \sum_{j=0}^{\infty} P[b(k) = j | a(k) = i] z^n$$
$$= G_H^i(z) \cdot G_L^{N-i}(z) \tag{8.28}$$

where f^i means the multiplication of i times.

The steady-state number of arrivals per time slot is then given by

$$\lambda_{tot} = N[\pi_H G_H'(z)|_{z=1} + \pi_L G_L'(z)|_{z=1}] \tag{8.29}$$

8.2.3 $\Sigma MMBP/D/1$

In this section, we take a look at a queueing system where an infinite waiting queue is fed by a group of N identical MMBP sources and served by a deterministic server. Without loss of generality, we assume that the width of a time slot is equal to the constant service time $(1/\mu)$ and is equal to 1 unit of measure.

Having obtained the correlation expressions (8.27) and (8.28) for those random variables presented in the last section, we are now in a position to examine the evolution of queue length in slot $(k+1)$, that is $q(k+1)$. Firstly, let us define the z-transform of the joint probability $P[q(k+1) = j, a(k+1) = i]$:

$$Q_i^{k+1}(z) = \sum_{j=0}^{\infty} P[q(k+1) = j, a(k+1) = i] z^j$$
$$= \sum_{j=0}^{\infty} \pi_{ji}^{k+1} z^j \tag{8.30}$$

In matrix form, we have

$$\tilde{Q}^{k+1}(z) = \begin{bmatrix} Q_0^{k+1}(z) \\ Q_N^{k+1}(z) \end{bmatrix} = \begin{bmatrix} \pi_{00}^{k+1}z^o & \pi_{j0}^{k+1}z^j \\ \pi_{01}^{k+1}z^o & \pi_{j1}^{k+1}z^j \\ \vdots & \vdots \\ \pi_{0N}^{k+1}z^o & \pi_{jN}^{k+1}z^j \end{bmatrix}$$

According to the model described, the queue will evolve according to

$$q(k+1) = [q(k)-1]^+ + b(k+1) \tag{8.31}$$

where $[a]^+$ denotes a or zero. If we examine $Q_0^{k+1}(z)$ and expand it in terms of those quantities in slot k using Equation (8.31), we will arrive at the following expression after some lengthy algebraic manipulation:

$$Q_0^{k+1}(z) = \sum_{i=0}^{N} a_{i0} B_0(z) \left(\frac{Q_i^k(z)}{z} + \frac{\pi_{0i}^k z - \pi_{0i}^k}{z} \right) \tag{8.32}$$

where $B_0(z)$ is as defined in Equation (8.28). Similarly, we can derive similar expressions for other terms $Q_1^{k+1}(z), \ldots, Q_N^{k+1}(z)$. If we assume that

$$\begin{aligned} \rho &= \lambda_{tot}/\mu \\ &= N(\pi_H G'_H(z)|_{z=1} + \pi_L G'_L(z)|_{z=1}) \\ &< 1 \end{aligned}$$

then the queue system will reach an equilibrium and we have

$$\lim_{k \to \infty} \tilde{Q}^{k+1}(z) = \lim_{k \to \infty} \tilde{Q}^k(z) = \lim_{k \to \infty} \tilde{Q}(z)$$

Putting all terms in matrix form, we have the following matrix equation:

$$\tilde{Q}(z) = \left(\tilde{I} - \frac{1}{z} \tilde{\Lambda}_B \tilde{A}^T \right)^{-1} \cdot \frac{1}{z} \tilde{\Lambda}_B \tilde{A}^T \cdot \tilde{\Phi} \tag{8.33}$$

where

$$\begin{aligned} \tilde{\Lambda}_B &= diag(B_0(z), \ldots, B_N(z)) \\ &= \begin{pmatrix} B_0(z) & 0 & \cdots & 0 \\ \vdots & B_1(z) & & \\ \vdots & & \ddots & \\ 0 & 0 & & B_N(z) \end{pmatrix} \end{aligned} \tag{8.34}$$

$B_i(z)$ is the z-transform defined in Equation (8.28) and $\tilde{A}$ is the matrix whose rows are defined as in Equation (8.27) and

$$\tilde{\Phi} = \begin{pmatrix} \pi_{00}z - \pi_{00} \\ \pi_{01}z - \pi_{01} \\ \pi_{0n}z - \pi_{0N} \end{pmatrix} \tag{8.35}$$

Some interesting results can be observed from this expression. If we set $z = 1$ in Equation (8.33), we have a set of linear equations as follows:

$$(z\tilde{I} - \tilde{\Lambda}_B\tilde{A}^T)\cdot\tilde{Q}(z) = \tilde{\Lambda}_B\tilde{A}^T\cdot\tilde{\Phi}$$

$$\left\{ \begin{pmatrix} 1 & 0 & \cdots & 0 \\ 0 & 1 & & \\ \vdots & & \ddots & 0 \\ 0 & & 0 & 1 \end{pmatrix} - \begin{pmatrix} a_{00} & a_{10} & \cdots & a_{N0} \\ a_{01} & a_{11} & & a_{N1} \\ & & \ddots & \\ a_{0N} & a_{1N} & & a_{NN} \end{pmatrix} \right\} \cdot \begin{pmatrix} A_0 \\ A_1 \\ \vdots \\ A_N \end{pmatrix} = \begin{pmatrix} 0 \\ 0 \\ \vdots \\ 0 \end{pmatrix}$$

Note that A_i is the steady-state probability of having i sources in H-state and

$$Q_i(z)|_{z=1} = A_i$$
$$= \pi_{0i} + \pi_{1i} + \ldots + \pi_{Ni}$$

This is exactly the set of steady-state balance equations that governed the behaviour of the set of N two-state Markov chains by solving the equation:

$$\tilde{\pi}\tilde{A} = \tilde{\pi}$$

8.2.4 Queue Length Solution

Earlier, we obtained an explicit expression (8.33) for the joint probability of queue length and traffic source. However, we notice that the z-transform of queue length (q) is given by

$$q(z) = \sum_{j=0}^{\infty} P[q = j]z^j$$

$$= \sum_{i=0}^{N} Q_i(z) = \tilde{e}\tilde{Q}(z) \tag{8.36}$$

From Equation (8.33) we have

$$\tilde{Q}(z) = \left(\tilde{I} - \frac{1}{z} \tilde{\Lambda}_B \tilde{A}^T \right)^{-1} \cdot \frac{1}{z} \tilde{\Lambda}_B \tilde{A}^T \cdot \tilde{\Phi}$$

$$= \sum_{j=0}^{\infty} z^{-j} (\tilde{\Lambda}_B \tilde{A}^T)^{j+1} z^{-1} \tilde{\Phi}$$

$$= \sum_{j=0}^{\infty} z^{-j-1} (\tilde{\Lambda}_B \tilde{A}^T)^{j+1} \tilde{\Phi} \qquad (8.37)$$

From the section 'Matrix Operation' in Chapter 1, we know that $\tilde{\Lambda}_B \tilde{A}^T$ can be expressed in terms of three matrices as

$$\tilde{\Lambda}_B \tilde{A}^T = \tilde{H}(z) \Lambda(z) \tilde{H}^{-1}(z)$$

$$= \sum_{i=0}^{N} \lambda_i \tilde{g}_i \tilde{h}_i \qquad (8.38)$$

where $\tilde{H}(z)$ is the eigenvector matrix of $\tilde{\Lambda}_B \tilde{A}^T$ and $\tilde{\Lambda}$ is a diagonal matrix consisting of the corresponding eigenvalues of $\tilde{\Lambda}_B \tilde{A}^T$. $\tilde{g}_i$ is the column vector for $\tilde{H}(z)$ and $\tilde{h}_i$ the row vector for $\tilde{H}^{-1}(z)$.

Substituting into Equation (8.37), we have

$$\tilde{Q}(z) = \sum_{j=0}^{\infty} z^{-j-1} (\tilde{H}(z) \tilde{\Lambda}(z) \tilde{H}^{-1})^{j+1} \tilde{\Phi}$$

$$= \sum_{j=0}^{\infty} z^{-j-1} \left(\sum_{i=0}^{N} \lambda_i^{j+1} \tilde{g}_h \tilde{h}_i \right) \tilde{\Phi}$$

$$= \sum_{j=0}^{\infty} \frac{\lambda_i}{z - \lambda_i} \tilde{g}_i \tilde{h}_i \tilde{\Phi} \qquad (8.39)$$

To find the average queue length, let us take the derivative of $\tilde{e} \tilde{Q}(z)$ when $z = 1$:

$$E[q] = \tilde{e} \tilde{Q}'(z)|_{z=1}$$

$$= \tilde{e} (\tilde{I} - \tilde{A}^T)^{-1} \tilde{A}^T \tilde{\Phi}' \qquad (8.40)$$

where

$$\tilde{\Phi}' = (\pi_{00}, \ldots, \pi_{0N})^T.$$

8.2.5   Initial Conditions

To obtain queue length distribution, a group of initial conditions (initial condition vector Φ') must first be determined. We define the system characteristic function as $\Pi_i[z - \lambda_i(z)] = 0$. The roots of each equation (for each λ_i) must be one or more of the following three types:

(1) Vanishing roots : real roots and $|z| \leq 1$,
(2) Non-vanishing roots : real roots and $|z| > 1$,
(3) Ignored roots :complex roots.

Since $\tilde{Q}(z)$ must be finite in $|z| \leq 1$, only vanishing roots are used for obtaining initial conditions. For every vanishing root and for each given i, we can get an initial condition equation $\tilde{h}_i(z)\Phi' = 0$. Since the Markov chain we discussed is irreducible, it must have a root equal to 1, that is $\lambda_0(1) = 1$ must be true. For this root, the initial condition equation is obtained by taking the limit of $\tilde{e}\tilde{Q}(z)$ in Equation (8.37) as z goes to 1, leading to

$$eg_0(1)h_0\Phi' = 1 - \lambda_0'(1) = 1 - \rho \qquad (8.41)$$

This equation is used to solve the initial condition vector Φ'.

8.3   MARKOV-MODULATED FLUID FLOW

Fluid-flow modelling refers to continuous-space queues with continuous inter-arrival times. In this type of modelling, we no longer treat the arrivals and departures as discrete units. The arrival process is deemed as a source of fluid flow generating a continuous stream of arrivals that is characterized by a flow rate, and departure as a continuous depletion of the waiting queue of a constant rate.

Fluid-flow models are appropriate approximations to situations where the number of arrivals is relatively large so that an individual unit is by itself of little significance. The effect is that an individual unit has only an infinitesimal effect on the flow – like a droplet of water in a flow.

To adopt this model, the arrival traffic has to be heavy ($\rho > 90\%$) and hence the waiting times in a queue are sufficiently large compared to the service time. In this section, the arrival process is governed by a continuous-time Markov process, and hence the term Markov-modulated fluid flow.

8.3.1   Model and Queue Length Analysis

As a teaching example, we consider a queueing system whose continuous fluid arrival process is governed by a three-state Markov chain, as shown in Figure 8.7.

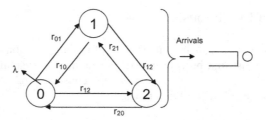

Figure 8.7   A Markov-modulated fluid model

When the Markov chain is state 0, customers arrive at a rate of λ and there are no arrivals in the other two states. If the depletion rate of the continuous-state queue is μ, then:

a) The queue length (q) grows at a rate of $(\lambda - \mu)$ when the arrival process is in state 0, and
b) The queue length decreases at a rate of μ in the other two states until the queue is empty.

Let $q_i(t, x)$, $i \in \{0, 1, 2\}$) and $x \geq 0$ be the probability that at time t the arrival process is in state i and the queue length does not exceed x. If we consider the probability change in an infinitesimal period Δt, then we have the following expressions:

$$q_0(t + \Delta t, x) = \{1 - (r_{01} + r_{02})\Delta t\}q_0(t, x - (\lambda - \mu)\Delta t) + r_{10}\Delta t q_1(t, x) \\ + r_{20}\Delta t q_2(t, x)$$

$$q_1(t + \Delta t, x) = r_{01}\Delta t q_0(t, x) + \{1 - (r_{10} + r_{12})\Delta t\}q_1(t, x + \mu)\Delta t) + r_{21}\Delta t q_2(t, x)$$

$$q_2(t + \Delta t, x) = r_{02}\Delta t q_0(t, x) + r_{12}\Delta t q_1(t, x) + \{1 - (r_{21} + r_{20})\Delta t\}q_2(t, x + \mu)\Delta t)$$

Rearranging terms, dividing both sides by Δt and letting Δt approach 0, we have the following set of partial differential equations:

$$\frac{\partial}{\partial t}q_0(t, x) + (\lambda - \mu)\frac{\partial}{\partial x}q_0(t, x) = -(r_{01} + r_{02})q_0(t, x) + r_{10}q_1(t, x) + r_{20}q_2(t, x)$$

$$\frac{\partial}{\partial t}q_1(t, x) - \mu\frac{\partial}{\partial x}q_1(t, x) = r_{01}q_0(t, x) - (r_{10} + r_{12})q_1(t, x) + r_{21}q_2(t, x)$$

$$\frac{\partial}{\partial t}q_2(t, x) - \mu\frac{\partial}{\partial x}q_2(t, x) = r_{02}q_0(t, x) + r_{12}q_1(t, x) - (r_{21} + r_{20})q_2(t, x)$$

Since we are interested in the steady-state behaviour of a queueing system, we set

$$\frac{\partial}{\partial t} q_i(t, x) = 0$$

hence we can drop the time parameter t and we have

$$(\lambda - \mu) \frac{d}{dx} q_0(t, x) = -(r_{01} + r_{02}) q_0(t, x) + r_{10} q_1(t, x) + r_{20} q_2(t, x) \quad (8.42)$$

$$-\mu \frac{\partial}{\partial x} q_1(t, x) = r_{01} q_0(t, x) - (r_{10} + r_{12}) q_1(t, x) + r_{21} q_2(t, x) \quad (8.43)$$

$$-\mu \frac{\partial}{\partial x} q_2(t, x) = r_{02} q_0(t, x) + r_{12} q_1(t, x) - (r_{21} + r_{20}) q_2(t, x) \quad (8.44)$$

The $q_i(x)$ is the steady-state probability that the arrival process is in state i and the queue length does not exceed x. In matrix form, we have

$$\tilde{D} \frac{d}{dx} \tilde{Q}(x) = \tilde{M} \tilde{Q}(x) \quad (8.45)$$

where

$$\tilde{Q}(x) = (q_0(x), q_1(x), q_2(x))^T$$

$$\tilde{D} = diag[(\lambda - \mu), -\mu, -\mu]$$

$$\tilde{M} = \begin{pmatrix} -(r_{01} + r_{02}) & r_{10} & r_{20} \\ r_{01} & -(r_{10} + r_{12}) & r_{21} \\ r_{02} & r_{12} & -(r_{21} + r_{20}) \end{pmatrix}$$

From the section 'Matrix Calculus' in Chapter 1, we see that the general solution of Equation (8.45) is

$$\tilde{Q}(x) = Q(\infty) + \sum a_i g_i e^{z_i x} \quad (8.46)$$

where the z_i's are eigenvalues of $\tilde{D}^{-1} \tilde{M}$ in the left-half complex plane (i.e. $Re(z_i) > 0$) and g_i are eigenvectors satisfying

$$z_i g_i = \tilde{D}^{-1} \tilde{M}$$

The coefficient a_i can be determined by boundary conditions. We see from the proceeding analysis that the fluid flow model has transformed the queueing problem into a set of differential equations instead of the usual linear equations.

8.3.2   Applications of Fluid Flow Model to ATM

The Markov-modulated fluid flow model is a suitable model for analysing traffic presented to a switch in an ATM network. In ATM, packets are fixed-sized cells of 53 bytes in length and the transmission speeds are probably of the order of a few gigabits per second, hence the transmission time of an individual cell is like a drop of water in the fluid flow model and so its effect is infinitesimal.

Another important advantage of the fluid-flow model is its ease of simulation. As discussed earlier, the transmission time of a cell is on a very fine scale, hence the simulation of cell-arrival events would consume vast CPU and possible memory resources, even for a simulation of a few minutes. A statistically meaningful simulation may not be feasible. In contrast, a fluid-flow model simulation deals with the rate of arrivals and rate of transmission and can be simulated over much longer time periods.

8.4   NETWORK CALCULUS

In this section, we are presenting an introductory overview of a new paradigm of queueing analysis called Network Calculus, which analyses queues from a very different perspective from that of classical queueing theory. For more detail, students should refer to the references.

Network Calculus is a new methodology that works with deterministic bounds on the performance measures of a queue. It was pioneered by Rene L. Cruz (Cruz 1991) for analysing delay and buffering requirements of network elements, and further developed by Jean-Yves Boudec (Boudec and Thiran 2002) and Cheng-Shang Chang (Chang 2000). It is a form of deterministic queueing theory for the reasons explained below.

In classical queueing theory, the traffic that enters a queueing system (or queueing network) is always characterized by a stochastic process; notably Poisson or other independent arrival processes. The associated calculus is then applied to derive the performance measures. The calculus here refers to the stochastic and mathematical theories that are associated with the arrival and service processes. The queueing results are often complex or the exact analysis intractable for realistic models. In addition, the traffic arrival processes are often correlated and hence render the Poisson or other simple arrival model unrealistic.

However, in Network Calculus, the arrival traffic is assumed to be 'unknown' but only satisfies certain regularity constraints, such as that the amount of work brought along by the arriving customers within a time interval is less than a value that depends on the length of that interval. The associated calculus is then developed to derive the deterministic bounds on the performance guarantees. Hence, Network Calculus is also referred to as deterministic queueing.

The main difference between these two approaches – classical queueing theory and Network Calculus – is that the former permits us to make statistical predictions of performance measures whereas the later establishes deterministic bounds on performance guarantees. For example, classical queueing theory may predict that 90% of the time a customer would wait less than T seconds before they receive their service, whereas the network calculus can ascertain that the time taken to go through a queue is less than Y.

8.4.1   System Description

The main idea of the Network Calculus is analogous to that of the classical system theory that has been long applied to electrical circuits. Given the input function of a signal in classical system theory, the output can be derived from the convolution of the input function and the impulse response function of the system.

Similarly, we can consider a queueing system (or queueing network) as a system. If we are able to put a constraint curve on the arrival process, then we should be able to establish certain bounds on the output process by applying certain calculus to both the arrival constraint curve and the 'characteristic function' of the queueing system. Figure 8.8 shows the schematic system view of a queueing system.

In contrast to the conventional approach of characterizing the arrival probabilistically using a stochastic process that counts the number of arrivals, we focus our attention on the cumulative amount of work brought along to the queueing system by customers. We describe the arrival process by a cumulative function $A(t)$, defined as the total amount of work brought to the queueing system by customers in time interval $[0, t]$. Without loss of generality, we take $A(t) = 0\ for\ t < 0$; in other word we assume that we begin with an empty system.

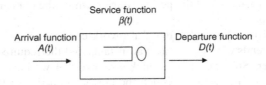

Figure 8.8   Schematic system view of a queue

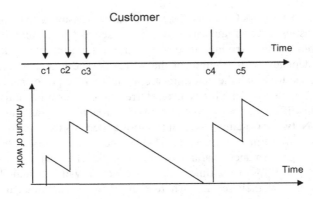

Figure 8.9   Arrivals and $X(t)$ content

Note that $A(t)$ is the sum of the service times required by those arriving customers in the time interval $[0, t]$.

Similarly, instead of counting the number of departing customers from a queueing system, we define an accumulative function $D(t)$ as the total amount of work completed by the queueing system in the time interval $[0, t]$, in other words, the completed services of those departed customers. Note that $D(t)$ is the cumulative amount of service time dispensed by the server in the time interval $[0, t]$, or the so-called finished work. Obviously:

$$D(t) \le A(t) \quad \text{and} \quad D(t) = 0 \quad for \quad t < 0 \tag{8.47}$$

At the same time, we modify the usual description of system queue length. The system queue length $X(t)$ here refers to the total amount of service times, so-called unfinished work, required by those customers in the system, rather than the number of customers in the system.

Figure 8.9 shows a sample path of $X(t)$ and its relationships to the arrivals. A new arrival of a customer causes the $X(t)$ to increase by the amount of work brought along by this customer. If we assume that the server is serving customers at a constant rate C, then the waiting queue is depleted at rate C when it is non-empty. The time periods during which the system content is non-empty are called busy periods, and the periods during which the content is empty are called idle periods.

It is obvious from Figure 8.9 that for a work-conserving system, the evolution of $X(t)$ depends only on the arrival instants and the required service time of that customer. Students should quickly realize that we have not made any assumptions about the ways in which the customers arrive at the queue, nor the service process. We only assume that the server dispenses his/her service at a constant rate C.

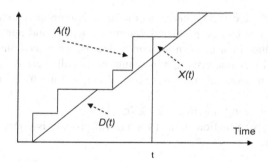

Figure 8.10 Sample path of $A(t)$ and $D(t)$

Assuming that the waiting queue is infinite, hence there is no overflow during [0, t], the system queue length at time t is given by

$$X(t) = A(t) - A(s) + X(s) - C \cdot (t - s) \quad 0 < s < t \tag{8.48}$$

$A(t) - A(s)$ is the amount of work brought along by those customers in $[s, t]$, and $C(t - s)$ is the amount of service that has been dispensed by the servers. We assume here that A is left-continuous. To cater for the situation where there is an arrival at $t = 0$ as well as the situation where the work arrival over [0, t] had a rate less than C so that the queue never built up and $X(t) = 0$, we have

$$X(t) = \sup_{s \leq t} \{ A(t) - A(s) - C \cdot (t - s) \} \tag{8.49}$$

The short-hand notation 'sup' refers to 'supremum', which we will explain in a later section. Similarly, we can derive an expression for the departure process:

$$\begin{aligned} D(t) &= A(t) - X(t) \\ &= A(t) - \sup_{s \leq t} \{ A(t) - A(s) - C \cdot (t - s) \} \\ &= \inf_{0^- \leq s \leq t} \{ A(s) + C \cdot (t - s) \} \end{aligned} \tag{8.50}$$

where 'inf' refers to 'infimum', also to be explained later. Figure 8.10 shows a sample path of $A(t)$ and $D(t)$.

8.4.2 Input Traffic Characterization – Arrival Curve

From the system description in Section 8.4.1, we see that we are moving away from a discrete space description of the arrival and system content into the continuous-time domain and hence a kind of Fluid Model.

We need to place certain constraints on the arrival process so that the output process of a queue can be predicted deterministically; and hence put bounds on its performance measures. In network calculus, we place constraints on the arrival process by characterizing it by using the so-called arrival curve. We say that the arrival process has $\alpha(t)$ as an arrival curve if the following are true:

a) $\alpha(t)$ is an increasing function for $t \geq 0$;
b) The cumulative function $A(t)$ of the arrival process is constrained by $\alpha(t)$ for $s \leq t$, e.g.:

$$A(t) - A(s) \leq \alpha(t - s)$$

For example, if $\alpha(t) = rt$, then the constraint means that during any time interval T_0, the amount of work that goes into the queue is limited by rT_0. This kind of arrival process is also known as peak-rate limited. In the domain of data communication, this peak-rate limited arrival may occur when data are arriving on a link whose bit rate is limited by r bps. This type of data flow is often called a 'constant bit rate (CBR)' flow.

Example 8.3

In Integrated Services Networks, the traffic specification often defines an arrival constraint curve $\alpha(t) = \min(M + pt, rt + b)$ for the flow, where M is the maximum packet size, p the peak rate, b the burst tolerance, and r the sustainable rate. The arrival curve is as shown:

8.4.3   System Characterization – Service Curve

In deriving the expression (8.50), we assume that the system is serving customers at a constant rate C. Let us generalize this and assume the service is constrained by a function $\beta(t)$. Then, it can be rewritten as

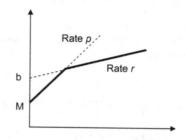

Figure 8.11   Sample Arrival Curve

$$D(t) = \inf_{0^- \leq s \leq t} \{A(s) + \beta(t-s)\}$$

$$= (A \otimes \beta)(t) \tag{8.51}$$

The $\otimes$ notation is the so-called convolution operation in min-plus algebra. The expression resembles the convolution of the input signal function and the impulse function to yield the output process in system theory, except the min-plus convolution operation here replaces multiplication with addition, and addition with the computation of infimum.

This leads us to the definition of the service curve. We say that a system offers to the input process a service curve β if and only if

$$D(t) = (A \otimes \beta)(t) \tag{8.52}$$

From Equation (8.51), we see that given the characterization of arrival process in terms of arrival curve and the characterization of service process in terms of service curve, the departing process of a queue can be calculated deterministically. This is the essence of network calculus.

8.4.4   Min-Plus Algebra

In previous sections, we introduced the notions of infimum and supremum. We shall clarify these terms here. Students can refer to (Boudec and Thiran 2002) for more detail.

Conceptually, infimum and supremum are similar to the minimum and maximum of a set respectively with some minor differences. Consider a non-empty set S $\{s \in S\}$; it is said to be bounded from below if there is a number N_1 such that $s \geq N_1$. That number N_1 is the infimum of S. If N_1 happens to be an element in set S and is smaller than any other elements in the set, then is called the minimum of S. For example, the closed and open interval of [a, b] and (a, b) have the same infimum, but (a, b) does not have a minimum.

Similarly, S is said to be bounded from above if there is a number N_2 such that $S \leq N_2$, then N_2 is the supremum of S. If it happens to be an element in the set and is larger than any other elements in the set, it is called the maximum of S. For example, sup [4,5,6] = max [4,5,6] = 6.

In conventional linear system theory, the convolution between two functions is defined as

$$(f \otimes g)(t) = \int_{-\infty}^{\infty} f(t-s)g(s)ds$$

However, in min-plus algebra, the usual addition operator is replaced by the inf (or min) operator and the multiplication operator by the addition operator. Hence, the convolution of two functions is given as

$$(f \otimes g)(t) = \inf_{0 \le s \le t} [f(s) + g(t-s)] \qquad (8.53)$$

Similar to the conventional convolution, the min-plus convolution is associative, commutative and distributive as

$$(f \otimes g) \otimes h = f \otimes (g \otimes h) \qquad \text{Associativity}$$

$$f \otimes g = g = g \otimes f \qquad \text{Commutativity}$$

$$[\inf(f, g)] \otimes h = \inf(f \otimes h, g \otimes h) \qquad \text{Distributivity}$$

9

Flow and Congestion Control

Today, we often experience congestion in modern cities, in terms of long queues of cars, buses and people waiting for services. Congestion is usually caused by unpredictable events. Although the daily rush hour is semi-predictable, congestion can also be random due to breakdowns and accidents that cause delays and long queues stretching for long distances. Therefore, a control mechanism might be implemented, such as using traffic lights to delay access to a junction or by restricting access to crowded areas in central business districts through tolls and congestion charges.

In general, the communication network is similar to our everyday experiences and consists of finite resources, such as bandwidth, buffer capacity for packets and transmission capacity that depend on traffic control mechanisms to alleviate congestion problems.

There are specific requirement guarantees, referred to as Quality of Service (QoS), specified for each traffic flow. When the offered traffic load from the user to the network exceeds the design limit and the required network resources, congestion will often occur. In this section, we introduce several specific examples, where queueing theory has been successfully applied to study the behaviour of the network under these extreme conditions. We describe key parameters of performances, i.e. throughput γ and time delay $E[T]$. It is important to examine the basic functioning of the network, as shown in Figure 9.1.

9.1 INTRODUCTION

The topic of congestion control has a long history (Schwartz 1987). A simple mechanism for preventing congestion is flow control, which involves

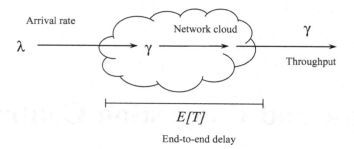

Figure 9.1 Flow control design based on queueing networks

regulating the arrival rate at which the traffic enters the network cloud, shown in Figure 9.1. In general, the *network layer* carries out *routing* functions in the network and provides adequate *flow control* to ensure timely delivery of packets from source to destination. The flow control protocols studied here are related to a packet-switched network, and the main functions of flow control are:

- Prevention of throughput and response time degradation, and loss of efficiency due to network and user overload;
- Deadlock avoidance;
- Fair allocation of resources among competing users;
- Speed matching between the network and attached users.

One simple taxonomy (Yang and Reddy 1995) that describes this well is the defined congestion control as an open and closed congestion control mechanism. The open-loop flow control mechanism is characterized by having no feedback between the source and the destination node. This control mechanism requires the allocation of resources with necessary prior reservation beforehand. However, the open-loop flow control has inherent problems with maximizing the utilization of the network resources. Resource allocation is made at connection setup using a CAC (Connection Admission Control) and this allocation is made using information that is already 'old news' during the lifetime of the connection. Often this process is inefficient and therefore results in over-allocation of resources.

For example, in ATM networks, the open-loop flow control is used by CBR (Constant Bit Rate), VBR (Variable Bit Rate) and UBR (Unspecified Bit Rate) traffic. On the other hand, the closed-loop flow control mechanism is characterized by the ability of the network to report pending network congestion back to the source node. This information is then used by the source node in various ways to adapt its activity, for example, by slowing down its rate depending on existing network conditions.

In this chapter, we examine the sliding window and rate-based flow control mechanism, which is a mathematical abstraction of a closed-loop flow control mechanism, where we can apply the earlier results provided in the chapters analyzing the performance of the networks. *Flow control* regulates the flow of data or information between two points (i.e. the receiver controls the rate at which it receives data). *Congestion control* maintains the number of packets or calls within the network to some level or region of the network so that the delay does not increase further. Performance relating to network throughput and time delay needs to be evaluated quantitatively to help us design an optimum flow control. Analysis based on a network of queues suffers from a high computational burden. A more compact measure, known as the *power*, is defined as the ratio of throughput and delay performances.

9.2   QUALITY OF SERVICE

In Chapter 8 we briefly discussed the traffic models that are used in ATM networks. An ATM network is primarily used in the backbone network. The traffic in ATM is divided into five classes of service categories (McDysan and Spohn 1995). The main characteristics of each service and the application is explained below and summarized in Table 9.1.

Table 9.1   Main characteristics of each service and their application

Types Of Services	Feedback	Examples
CBR	No	Constant Rate Video-on-Demand
RT-VBR	No	Variable Rate Video-on-Demand
NRT-VBR	No	Transaction Processing
UBR	No	Image Transfer
ABR	Yes	Critical Data Transfer

- *Constant Bit Rate (CBR)*
  The CBR service category is used for connections needing a static amount of bandwidth continuously throughout the connection. CBR is intended to support real-time applications, such as interactive video-on-demand and audio application.
- *Real-Time Variable Bit Rate (RT-VBR)*
  RT-VBR is intended for real-time applications, such as video-on-demand and audio, which transmit at a variable rate over time.
- *Non-Real-Time Variable Bit Rate (NRT-VBR)*
  NRT-VBR is intended for non real-time bursty applications, such as transaction processing, which have variable cell rates.
- *Unspecified Bit Rate (UBR)*
  UBR is intended for non real-time bursty applications such as text, data and image transfer. The user sends whenever they want to, with no guarantees and cells may be dropped during congestion.

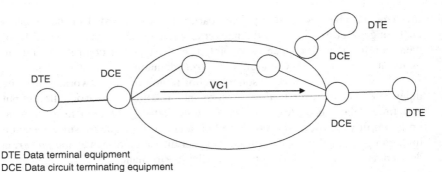

DTE Data terminal equipment
DCE Data circuit terminating equipment

Figure 9.2   Data network

- *Available Bit Rate (ABR)*
  ABR services use the capacity of the network when available and control
  the source rate by use of feedback. One major difference of ABR compared
  to the other four services is that it uses a closed-loop feedback to implement
  flow control. Some examples of ABR applications include critical data
  transfer and distributed file service. The ATM's ABR is a rate-based flow
  control mechanism that changes the source transmission rate according to
  explicit feedback from the network.

9.3   ANALYSIS OF SLIDING WINDOW FLOW CONTROL MECHANISMS

Many of the Internet's flow control mechanisms use a window in order to avoid
congestion. Figure 9.2 shows that the queueing model for the virtual traffic
VC1 can be a tandem queueing network. Furthermore, we can compare various
control mechanisms. Therefore, this abstraction is useful since it utilizes queueing networks to model a data network.

9.3.1   A Simple Virtual Circuit Model

A VC virtual circuit is a transmission path set up end-to-end, with different
user packets sharing the same path. Consider a VC, which has M store and
forward nodes from the source to the destination. It can be modelled as M
queues in series, as shown in Figure 9.3.

For the ith queue in the VC, the transmission rate or capacity is given as μ_i
packets per second. It is also common to assume the following:

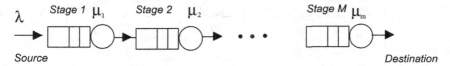

Figure 9.3 Simplified model for a single virtual circuit

(i) propagation delay is negligible, and the delay consists of queuing delay (waiting time) and transmission delay;
(ii) $1/\mu_i$ at the ith node represents the average time required to transmit a packet over its outgoing link;
(iii) independence and the packet lengths are exponentially distributed.

These assumptions make the analysis more tractable and lead to a *product form* closed queueing network.

9.3.2 Sliding Window Model

The analysis follows [Schwartz 1988] where the sliding window flow control protocol is often described in the following manner:

(a) Each packet is individually ACKnowledged, when it arrives at the destination node.
(b) The ACK, on arriving back at the source node, shifts the window forward by 1.
(c) Packets can be transmitted onto the VC as long as there are fewer than N (window size) packets along the VC.
(d) Arriving packets are assumed blocked and lost to the system if the window is depleted (i.e. N outstanding unacknowledged packets).

Suppose we neglect the delay incurred by acknowledgment returning from the destination end of the VC back to the source end. Hence we have a closed network, as shown in Figure 9.4 below:

When N packets are in transit along the VC, the bottom queue is empty and cannot serve (depleted window condition). Once one of the packets arrives at its destination, it appears in the $M + 1$ queue and the source can deliver the packet at a rate of λ.

Normally, we will find a trade-off between time delay and throughput. Firstly, we will introduce a theorem that is useful to analyse this sliding window queueing network, which is familiar to circuit analysis.

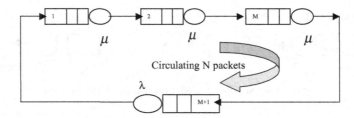

Figure 9.4 Sliding window control model (closed queueing network)

Theorem 9.1

For product form networks, any sub-network can be replaced by one composite queue with a state dependent service rate. By forming this network, the remaining queueing network will retains the exact statistical behaviour that is useful to analyse the network.

Example 9.1

By applying Norton's theorem to the sliding window flow control model in Figure 9.4, the upper queue is a generalized birth-death process with arrival rate λ and state dependent service rate u(n). The total number of packets in the network is N. The equilibrium probability when the state of the upper queue is n will be denoted by p_n.

$$\frac{p_n}{p_0} = \frac{\lambda^n}{\displaystyle\prod_{i=1}^{n} u(i)} \tag{9.1}$$

This follows from the *flow balance* condition which states that flow from state n to $n - 1 =$ flow from state $n - 1$ to n:

$$\Rightarrow p_n u(n) = p_{n-1} \lambda$$

The probability normalization condition also holds, i.e.:

$$\sum_{n=0}^{N} p_n = 1 \tag{9.2}$$

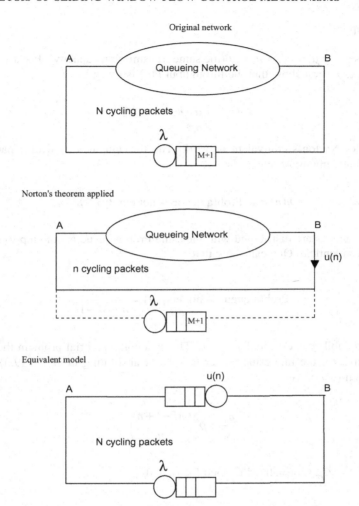

Figure 9.5 Norton aggregation or decomposition of queueing network

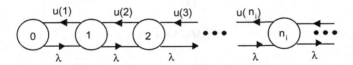

Figure 9.6 State transition diagram

Assumption

Suppose $\mu_1 = \mu_2 = \ldots = \mu_m = \mu$ (the same transmission capacity). For a simple network, we can show that the throughput $u(n)$ is given by

$$u(n) = n\mu / (n + (M - 1)) \qquad (9.3)$$

Consider Norton's equivalent network. The throughput $u(n)$ with n packets distributed among M queues is

$$u(n) = \mu \cdot \text{Prob(a queue is not empty)} \le \mu \qquad (9.4)$$

All the queues are distributed with the same Prob(a queue is not empty) since they are identical. One can show that

$$\text{Prob(a queue is not empty)} = \frac{n}{n + M - 1}$$

The probability is essentially $1 - p_0$. This is a combinatorial problem that we can leave for the interested reader to solve. Substituting Equation (9.3) into Equation (9.1) gives

$$\frac{p_n}{p_0} = \rho^n \binom{M - 1 + n}{n} \qquad (9.5)$$

where $\rho = \lambda/\mu$ (normalized applied load), and

$$\frac{1}{p_0} = \sum_{n=0}^{N} \rho^n \binom{M - 1 + n}{n} \qquad (9.6)$$

where

$$\binom{M - 1 + n}{n}$$

is the binomial coefficient, i.e. number of combination of $M - 1 + n$ quantities taking n at a time. The throughput (end-to-end) of the window flow controlled VC is averaged over all the N possible service rates:

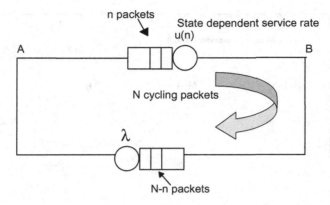

Figure 9.7 Norton's equivalent, cyclic queue network

$$\gamma = \sum_{n=1}^{N} u(n)p_n \qquad (9.7)$$

We need to apply Little's formula, which is a general result applicable to any queue. This states that if a queue has an average customer arrival rate of λ, then the average number of customers N in the queue and the average time spent by a customer in the queue will be related by $N = \lambda W$. By Little's formula, the end-to-end delay through the VC is

$$E[T] = \frac{\text{average number of packets}}{\text{end to end throughput}} = \frac{E[n]}{\gamma} = \frac{\sum_{n=1}^{N} np_n}{\gamma} \qquad (9.8)$$

Heavy traffic analysis

(i) Let $\lambda \gg \mu$, or $\lambda \to \infty$

We note that the upper queue is almost always at state N, i.e. $E(n) \approx N$:

$$\gamma = u(N) = \frac{N\mu}{M-1+N} \qquad (9.9)$$

and Little's theorem:

$$E[T] = \frac{N}{\gamma} = \frac{M-1+N}{\mu} \qquad (9.10)$$

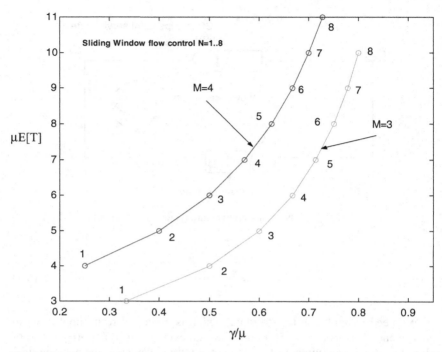

Figure 9.8 Delay throughput tradeoff curve for sliding window flow control, $M = 3$, 4 and $\lambda \to \infty$

By combining Equations (9.9) and (9.10), the time delay versus throughput tradeoffs characteristics are

$$\mu E[T] = \frac{M-1}{1-\dfrac{\gamma}{\mu}} \tag{9.11}$$

As $\gamma \to \mu$ (by increasing N), small changes in γ give rise to large changes in $E(T)$. Various criteria give the value of optimum N, i.e. $N = M - 1$

(ii) When $\lambda = \mu$, (load in the vicinity of the capacity of μ of each queue), we note that there are in effect $M + 1$ queues instead of M queues (see Figure 9.4):

$$\gamma = \frac{N\mu}{M+N} \tag{9.12}$$

and the new time delay expression:

$$\mu E[T] = \underbrace{\left(\frac{M}{M+1}\right)}_{\text{mod}ified\ to\ account\ for\ M+1\ queues} (M+N) \tag{9.13}$$

The new optimum value of N is

$$N = M$$

For the closed queuing network, the product form solution for this type of network is

$$p(\mathbf{n}) = \frac{1}{g(N, M)} \prod_{i=1}^{M} \left(\frac{\lambda_i}{\mu_i}\right)^{n_i} \tag{9.14}$$

$g(N, M)$ is called the *normalization constant* (statistical parameter of interest). Note that only 'relative' values of λ_i can be found and so $\rho_i = \lambda_i/\mu_i$ no longer represents the actual utilization of the ith queue.

Buzen's Algorithm (Convolution)

$$g(n, m) = g(n, m-1) + \rho_m g(n-1, m) \tag{9.15}$$

with initial starting condition:

$$g(n, 1) = \rho_1^n \quad n = 0, 1, 2, \ldots, N \tag{9.16}$$

$$g(0, m) = 1 \quad m = 1, 2, \ldots, M \tag{9.17}$$

where

$$\rho_m = \frac{\lambda_m}{\mu_m}$$

The important desired statistics is contained in the $g(N, M)$ normalization constant. $g(N, M)$ is also related to the moment generating function. Consider the number of packets in queueing station i to be equal to or exceed k packets. Then

$$\text{Prob}(n_i \geq k) = \rho_i^k g(N - k, M)/g(N, M) \tag{9.18}$$

Utilization of server: By letting $k = 1$ in equation (9.18),

$$\text{Prob}(n_i \geq 1) = 1 - \text{Prob}(n_i = 0):$$
$$= \rho_i g(N-1, M)/g(N, M) \qquad (9.19)$$

Throughput or the average number of packets serviced per unit time for the ith queue is given as

$$\gamma_i = \mu_i \text{Prob}(n_i \geq 1) = \mu_i \rho_i g(N-1, M)/g(N, M)$$
$$= \lambda_i g(N-1, M)/g(N, M)$$

The average number of packets in ith queue is

$$E[n_i] = \sum_{k=1}^{N} k \, \text{Prob}\{n_i = k\}$$

Now the marginal probability that queue i has k packets, i.e. $\text{Prob}(n_i = k)$:

$$\text{Prob}(n_i = k) = \text{Prob}(n_i \geq k) - \text{Prob}(n_i \geq k+1) \qquad (9.21)$$

Substitution for the above and simplifying gives

$$E[n_i] = \sum_{k=1}^{N} \rho_i^k g(N-k, M)/g(N, M) \qquad (9.22)$$

Example 9.2

Consider the following virtual circuit from node 1 to node 4 for packet-switching network transmission rates (μ, 2μ, 3μ packets/s) for the links, as indicated in Figure 9.9. Assume that the packet arrivals to this virtual circuit are a Poisson arrival process with a rate of λ packets/s:

(a) Show and explain how the sliding window flow control mechanism for packet-switched networks can be represented in terms of a network of $M/M/1$ queues.
(b) With a window size of $N = 1, \ldots, 5$ for the sliding window flow control mechanism for this virtual circuit, use Buzen's algorithm to find the end-to-end time delay and throughput for $\lambda \gg 3\mu$.

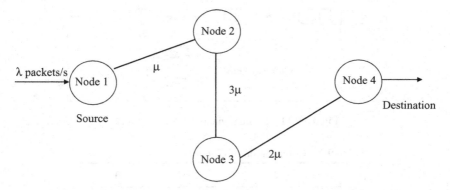

Figure 9.9   Packet-switching network transmission rates

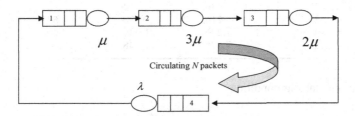

Figure 9.10   Typical closed network

Solution

a) Suppose we neglect the delay incurred by acknowledgement return from the destination end of the VC back to the source end. Hence we have a closed network, as shown Figure 9.10:

   Therefore N packets are in transit along VC and the bottom queue is empty and cannot serve (depleted window condition). Once one of the packets arrives at its destination, it appears in queue 4 and the source can deliver the packet at a rate of λ.

b) When $\lambda \gg 3\mu$, the sliding window can be modelled, as in Figure 9.11:

   N is the size of the window, while $M = 3$. Let $\lambda_1 = \lambda_2 = \lambda_3$ by flow conservation, and $\rho_2 = \rho_1/3$, $\rho_3 = \rho_1/2$. Now let us define $\rho_1 = 6$, so that $\rho_2 = 2$ and $\rho_3 = 3$. By using Buzen's algorithm:

$$g(n, m) = g(n, m - 1) + \rho_m g(n - 1, m)$$

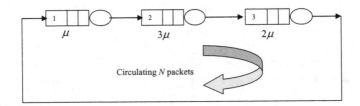

Figure 9.11   Sliding window closed network

Table 9.2   Computation of $G(n, m)$

ρ	6	2	3
n \ m	1	2	3
0	1	1	1
1	6	8	11
2	36	52	85
3	216	320	575
4	1,296	1,936	3,661
5	7,776	11,648	22,631

With initial starting condition:

$$g(n, 1) = \rho_1^n \quad n = 0, 1, 2, \ldots, N$$

$$g(0, m) = 1 \quad m = 1, 2, \ldots, M \text{ where } \rho_m = \frac{\lambda_m}{\mu_m}$$

One possible $G(n, m)$ is computed in Table 9.2 below:
The end-to-end throughput can be obtained from

$$\gamma_i = \mu_i \rho_i \frac{g(N-1, M)}{g(N, M)} = \lambda_i \frac{g(N-1, M)}{g(N, M)}$$

which simplifies to

$$= 6 \frac{g(N-1, 3)}{g(N, 3)}$$

The mean number of packets at each queue can be obtained by

$$E[n_i] = \sum_{k=1}^{N} \rho_i^k \frac{g(N-k, M)}{g(N, M)}$$

Table 9.3 Normalized end-to-end throughput and delay

N	1	2	3	4	5
γ/μ	0.545	0.776	0.887	0.942	0.971
μT	1.834	2.577	3.382	4.246	5.149

The delay over each of the virtual circuits can be determined by Little's formula:

$$T = \sum_{i=1}^{M} T_i = \sum_{i=1}^{3} \frac{E[n_i]}{\gamma_i} = \frac{\sum_{i=1}^{3} \sum_{k=1}^{N} \rho_i^k g(N-k, 3)/g(N, 3)}{\gamma}$$

When $\lambda \gg 3\mu$, it is obvious that it is not necessary to calculate the numerator term as it is

$$E[n] = N$$

the window size. So the normalized end-to-end throughput and delay can be calculated as a function of the different window size N, as shown in Table 9.3:

9.4 RATE BASED ADAPTIVE CONGESTION CONTROL

For the ATM network, the ABR service will systematically and dynamically allocate available bandwidths to users by controlling the rate of offered traffic through feedback. The aim of the ABR service is to support applications with vague requirements for throughput and delay. Due to these vague requirements, they are best expressed as ranges of acceptable values. The ABR service allows a user to specify the lower and upper bound on the bandwidth allotted upon connection. The control mechanism is based on feedback from the network to the traffic source.

The ability for the ABR service to work in a variety of environments, i.e. LANs, MANs or WANs, is particularly important when traffic is controlled using feedback. The motivation for feedback congestion control is based on adaptive control theory. A simple model may be derived to control the source rate $\lambda(t)$ to optimize a particular objective, such as the throughput or stability. By providing information regarding the buffer occupancy of the bottleneck queue, we can adjust the rate accordingly. The reader is advised to refer to advanced textbooks on the subject, such as (McDysan 2000).

References

Allen, A.O. *Probability, Statistics and Queueing Theory*. Academic Press Inc., 1978.

Anurag K.D. and Manjunath, J.K. *Communication Networking: An Analytical Approach*. Morgan Kaufmann Publishers, 2004.

Anick, D., Mitra D. and Sondhi, M.M. 'Stochastic Theory of a Data-Handling System with Multiple Sources'. *Bell Technical Journal* **61(8)**: 1982.

Bae, J.J, *et al.*, 'Analysis of Individual Packet Loss in a Finite Buffer Queue with Heterogeneous Markov Modulated Arrival Processes: A Study of Traffic Burstiness and Priority Packet Discarding'. *INFOCOM '92*, 219–230, 1992.

Baskett, F., Chandy, K.M., Muntz R.R. and Palacios-Gomez, F. 'Open, Closed and Mixed Networks of Queues with Different Classes of Customers'. *Journal of the ACM* **22(2)**: 248–260, 1975.

Bertsekas, D and Gallager, R. *Data Networks*, 2nd Edition. Prentice-Hall International Inc., 1992.

Blondia, C. and Casals, O. 'Statistical Multiplexing of VBR Sources: A Matrix-Analysis Approach'. *Performance Evaluation* **16**: 5–20, 1992.

Bolch, G., Greiner, S., de Meer, H. and Trivedi, K.S. *Queueing Networks and Markov Chains*. Wiley Interscience. 2006.

le Boudec, J-Y. and Thiran. P. *Network Calculus*. Springer Verlag, 2002.

Bruell, S.C. and Balbo, G. *Computational Algorithms for Closed Queueing Networks*. North-Holland, 1980.

Bruneel, H. 'Queueing Behavior of Statistical Multiplexers with Correlated Inputs'. *IEEE Trans. Commmunications*, **36**: 1339–1341, 1988.

Bruneel, H. and Byung G.K. *Discrete-Time Models for Communication Systems Including ATM*. Kluwer Academic Publishers, 1993.

Burke, P.J. 'The Output of a Queueing System'. *Operations Research* **4(6)**: 699–704, 1956.

Buzen, J.P. 'Computational Algorithms for Closed Queueing Networks with Exponential Servers'. *ACM* **16(9)**: 527–531, 1973.

Buzen, J.P. 'Operational Analysis: The Key to the New Generation of Performance Prediction Tools'. *Proceedings of the IEEE Compcon*, September 1976.

N. Cardwell, Savage, S. and Anderson, T. 'Modelling TCP Latency' *Proceedings of the IEEE Infocom*, 2000.

Chao, H.J. and Guo, X. *Quality of Service Control in High-Speed Networks*, 2002.

Chandy, K.M., Hergoy, U. and Woo, L. 'Parametric Analysis of Queueing Network Models'. *IBM Journal of R&D*, **19(1)**: 36–42, 1975.

Chang, C-S. *Performance Guarantees in Communication Networks*. Springer Verlag, 2000.

Chen, J.S.-C., Guerin, R. and Stern, T.E. 'Markov-Modulated Flow Model for the Output Queues of a Packet Switch'. *IEEE Trans. Communications*. **40(6)**: 1098–1110, 1992.

Chen, H. and Yao, D.D. *Fundamentals of Queueing Networks*, Springer-Verlag, 2001.

Cinlar, E. *Introduction to Stochastic Processes*. Prentice-Hall, 1975.

Cohen, J.W. *The Single Serve Queue*. North-Holland, 1982.

Diagle, J.N. and Langford, J.D. 'Models for Analysis of Packet Voice Comunications Systems'. *IEEE Journal Sel Areas Communications* **SAC-4(6)**: 1986.

Denning, P.J. and Buxen, J.P. 'The Operational Analysis of Queueing Network Models'. *Computer Surveys* **10(3)**: 1978.

Elwalid, A.I., Mitra, D. and Stern, T.E. 'Statistical Multiplexing of Markov Modulated Sources Theory and Computational Algorithms'. *ITC-13*, 495–500, 1991.

Elwalid, A.I. and Mitra, D. 'Fluid Models For The Analysis and Design of Statistical Multiplexing With Loss Priority On Multiple Classes of Bursty Traffic'. *IEEE INFOCOM '92*, 415–425, 1992.

Feller, W. *An Introduction to Probability Theory and Its Applications*, 3rd Edition, Volume I. John Wiley and Sons, 1968.

Firoiu, V. *et al*. 'Theory and Models for Internet Quality of Service'. *Proceedings of the IEEE*, May 2002.

Fischer, W. and Meier-Hellstern, K. 'The Markov-modulated Poisson Process (MMPP) Cookbook'. *Performance Evaluation* **18**: 147–171, 1993.

Fortz, B. and Thorup, M. 'Internet Traffic Engineering by Optimizing OSPF Weights'. *IEEE INFOCOM 2000*, 2000.

Frost, V.S. and Melamed, B. 'Traffic Modelling for Telecommunications Networks'. *IEEE Communications Magazine*, 70–81, 1994.

le Gall, D. 'MPEG: A Video Compression Standard for Multimedia Applications'. *Communications of the ACM* **34(4)**: 47–58, 1991.

Gebali, F. *Computer Communication Networks Analysis and Design*, Northstar Digital Design Inc, Victoria, BC 2005.

Gelenbe, E. and Pujolle, G. *Introduction to Queueing Networks*. John Wiley & Sons, 1987.

Gerla, M. and Kleinrock, L. 'Flow Control: A Comparative Survey.' *IEEE Trans. on Communications* **28(4)**: 553–574, 1980.

Ghanbari, M, and Hughes, C.J. 'Packing Coded Video Signals into ATM Cells'. *IEE/ACM Trans. on Networking* **1(5)**: 505–509, 1993.

Giambene, G. *Queueing Theory and Telecommunications*. Springer-Verlag, 2005.

Gordon, W.J and Newell, G.F. 'Closed Queueing Systems With Exponential Servers'. *Operations Research* **15(2)**: 254–265, 1967.

Gross, D and Harris, C.M. *Fundamentals of Queueing Theory*. John Wiley & Sons: New York, 1974.

Gunter, B., Greiner, S., de Meer, H., Trivedi, K.S. *Queueing Networks and Markov Chains.* John Wiley & Sons, 2006.

Habib, I.W. and Saadawi, T.N. 'Multimedia Traffic Characteristics in Broadband Networks'. *IEEE Communications Magazine* 48–54, July 1992.

Haight, F.A. *Applied Probability.* Plenum Press: New York and London, 1981.

Harrison, P.G. and Patel, N.M. *Performance Modelling of Communication Networks and Computer Architectures.* Addison-Wesley Publishing Company, 1993.

Hayes, J.F. *Modeling and Analysis of Computer Communications Networks.* Plenum Press: New York and London, 1984.

Heffes, H. and Lucantoni, D.M. 'A Markov Modulated Characterization of Packetized Voice and Data Traffic and Related Statistical Multiplexer Performance'. *IEEE Journal of Sel Areas in Communications* **SAC-4(6)**: 1986.

Horn, R.A. and Johson, C.R.. *Matrix Analysis.* Cambridge University Press, 1990.

Jackson, J.R. 'Networks of Waiting Lines'. *Operations Research* **5(4)**: 518–521, 1957.

Jackson, J.R. 'Jobshop-Like Queueing Systems'. *Management Science* **10(1)**: 131–142, 1963.

Jain, J.L., Mohanty, S.G. and Bohm, W. *A Course on Queueing Models.* Chapman & Hall, 2006.

Jaiswal, J.M. *Priority Queues.* Academic Press, 1968.

King, P J.B. *Computer and Communication Systems Performance Modelling.* Prentice-Hall, 1990.

Kleinrock, L. *Communication Nets: Stochastic Message Flow and Delay.* McGraw Hill: New York 1964, reprinted by Dover: New York 1972.

Kleinrock, L. *Queueing Systems,* Volume. 1. John Wiley & Sons: New York, 1975.

Kleinrock, L. 'Performance Evaluation of Distributed Computer-Communication Systems'. In: *Queueing Theory and Its Applications* (Boxma, O.J. and Syski, R. eds), North-Holland, 1988.

Kobayashi, H. *Modeling and Analysis: An Introduction to System Performance Evaluation Methodology.* Addison-Wesley 1978.

Lavenberg, S.S. and Reiser, M. 'Stationary State Probabilities at Arrival Instants of Closed Queueing Networks with Multiple Types of Customers'. *Journal of Applied Probability* **19**: 1048–1061, 1980.

Lee, Hyong W. and Mark, J.W. 'ATM Network Traffic Characterization Using Two Types of On-Off Sources' *IEEE INFOCOM '93:* 152–159, 1993.

Leduc, J-P. *Digital Moving Pictures Coding and Transmission on ATM.* Elsevier, 1994.

Li, S-Q. 'A General Solution Technique for Discrete Queueing Analysis of Multimedia Traffic on ATM'. *IEEE Trans. Communications* **39**: 115–1132, 1991.

Little, J. 'A Proof of the Queueing Formula $L = \lambda W$'. *Operations Research* **18**: 172–174, 1961.

Lucantoni, D.M., Meier-Hellstern, K.S. and Neuts, M.F. 'A Single Server Queue with Server Vacations and a Class of Non-renewal Arrival Process'. *Advances in Applied Probability* **22(2)**: 676–705, 1990.

Luhanga, M.L. 'A Fluid Approximation Model of an Integrated Packet Voice and Data Multiplexer'. *Proc of INFOCOM '88,* 687–692, 1988.

Magnus, J.R. and Neudecker, H. *Matrix Differential Calculus with Applications in Statistics and Econometrices.* John Wiley & Sons, 1988.

Martin, J. *Design of Real-Time Computer Systems*. Prentice Hall, 1967.

McDysan, D.E. and Spohn, D.L. *ATM: Theory and Application*. McGraw-Hill, 1995.

McDysan D.E., *QoS and Traffic Management in IP & ATM Network*. McGraw Hill, 2000.

Neuts, M.F. *Matrix-Geometric Solutions in Stochastic Models*. The John Hopkins University Press, 1981.

Neuts, M.F. *Structured Stochastic Matrices of M/G/1 Type and Their Applications*. Marcel Dekker, Inc., 1989.

Newman, P. 'ATM Technology for Corporate Networks'. *IEEE Communications Magazine*, 90–101, April 1992.

Nomura, M., *et al.* 'Basic Characteristics of Variable Rate Video Coding in ATM Environment'. *IEEE Journal Sel. in Communications* **7**(5): 752–760, 1989.

Mieghem, P.V. *Performance of Communications Networks and Systems*. Cambridge University Press, 2006.

Onvural, R.O. *Asynchronous Transfer Mode Networks Performance Issues*. Artech House, Inc., 1994.

Papoulis, A. *Probability, Random Variables, and Stochastic Processes*, 3rd Edition. McGraw-Hill, 1991.

Prabhu, N.U. and Zhu, Y. 'Markov-Modulated Queueing Systems'. *Queueing Systems* **5**: 215–246, 1989.

Ramamurthy, G. and Sengupta, B. 'Modeling and Analysis of a Variable Bit Rate Video Multiplexer'. *INFOCOM '92* 817–827, 1992.

Reiser, M. and Kobayashi, H. 'Numerical Solution of Semiclosed Exponential Server Queueing Networks'. *Proceedings of the 7th Asilomar Conference on Circuits Systems and Computers*, 308–312. November 1973.

Reiser, M. 'A Queueing Network Analysis of Computer Communication Networks with Window Flow Control'. *IEEE Trans. on Communications*, **27**(8): 1990–1209, 1979.

Reiser, M. and Lavenberg, S.S. Mean Value Analysis of Closed Multichain Queueing Networks'. *Journal of the ACM* **22**: 313–322, 1980.

Reiser, M. 'Performance Evaluation of Data Communication Systems'. *Proceedings of the IEEE* **70**(2): 171–195, 1982.

Ren, Q. and Kobayashi, H. 'Diffusion Process Approximations of a Statistical Multiplexer with Markov Modulated Bursty Traffic Sources'. *GLBECOM '94* **3**: 1100–1106, 1994.

Robertazzi, T.G. *Computer Networks and System: Queueing Theory and Performance Evaluation*, 3rd Edition, Springer-Verlag, 2000.

Robertazzi, T.G. *Planning Telecomunication Networks*. IEEE: Piscataway, NJ. 1999.

Ross, S.M. *Stochastic Processes*. John Wiley & Sons, 1983.

Sen, P., *et al.* 'Models for Packet Switching of Variable-Bit-Rate Video Sources'. *IEEE Journal of Sel Areas Communications* **7**(5): 865–869, 1989.

Schwartz, M. *Telecommunication Networks Protocol, Modeling and Analysis*. Addison-Wesley, 1987.

Schwartz, M. *Broadband Integrated Networks*. Prentice Hall, 1996.

Sevcik, K.C. and Mitrani I. 'The Distribution of Queueing Network States at Input and Output Instants'. *Journal of the ACM* **28**(2): 358–371, 1981.

Skelly, P.M., Schwartz, M. and Dixit, S. 'A Histogram-Based Model for Video Traffic Behavior in an ATM Multiplexer'. *IEEE/ACM Trans. on Networking* **1**(4): 446–459, 1993.

Spragin, J.D., Hammond, J.L. and Pawlikowski, K. *Telecommunications: Protocol and Design*. Addison-Wesley: Reading, MA., 1991.

Sohraby, K. 'On the Asymptotic Behavior of Heterogeneous Statistical Multiplexer With Applications'. *INFOCOM '92* 839–847, 1992.

Sriram, K. and Whitt, W. 'Characterizing Superposition Arrival Processes in Packet Multiplexers for Voice and Data'. *IEEE Journal of Sel. Area Communications* **SAC-4(6)**: 833–845, 1986.

Taylor, H.M. and Karlin, S. *An Introduction To Stochastic Modeling*. Academic Press, 1994.

Tian, N. and Zhe, G.Z. *Vacation Queueing Models: Theory and Applications*. Springer-Verlag, 2006.

Tijms, H.C. *Stochastic Models: An Algorithmic Approach*. John Wiley & Sons, 1994.

Van Dijk, N. *Queueing Networks and Product Forms A Systems Approach*. John Wiley & Sons, 1993.

Vesilo, R.A.. 'Long-Range Dependence of Markov Renewal Processes'. *Australian and New Zealand Journal of Statistics* **46(1)**: 155–171, 2004.

Wolff, R.W. 'Poisson Arrivals See Time Averages'. *Operations Research* **30**: 223–231, 1982.

Wright, S. 'Admission Control in Multi-service IP Networks: A Tutorial'. *IEEE Communications Surveys and Tutorials*, July 2007.

Xiong, Y. and Bruneel, H. 'A Tight Upper Bound for the Tail Distribution of the Buffer Contents in Statistical Multiplexers with Heterogeneous MMBP Traffic Sources'. *GLOBECOM '93* 767–771, 1993.

Yang, C. and Reddy, A. 'A Taxonomy for Congestion Control Algorithms in Packet Switching Network', *IEEE Networks*, July/August 1995.

Ye, J. and Li, S-Q. 'Analysis of Multi-Media Traffic Queues with Finite Buffer and Overload Control – Part I: Algorithm'. *INFOCOM '91*, 464–1474, 1991.

Ye, J. and Li, S-Q. 'Analysis of Multi-Media Traffic Queues with Finite Buffer and Overload Control – Part II: Applications'. *INFOCOM '92*, 1464–1474, 1992.

Index

Note: page numbers in italics refer to figures and diagrams